Managing Editor
Karen Goldfluss, M.S. Ed.

Editor-in-Chief
Sharon Coan, M.S. Ed.

Cover Artist
Barb Lorseyedi

Art Coordinator
Kevin Barnes

Art Director
CJae Froshay

Imaging
James Edward Grace

Product Manager
Phil Garcia

Publisher
Mary D. Smith, M.S. Ed.

Practice Makes Perfect

Fractions

GRADE 4

Authors

Teacher Created Resources Staff

Teacher
Created
Resources

Teacher Created Resources, Inc.
6421 Industry Way
Westminster, CA 92683
www.teachercreated.com

ISBN-0-7439-3325-7

©2002 Teacher Created Resources, Inc.

Reprinted, 2006
Made in U.S.A.

Table of Contents

Introduction

The old adage "practice makes perfect" can really hold true for your child and his or her education. The more practice and exposure your child has with concepts being taught in school, the more success he or she is likely to find. For many parents, knowing how to help your children can be frustrating because the resources may not be readily available. As a parent it is also difficult to know where to focus your efforts so that the extra practice your child receives at home supports what he or she is learning in school.

This book has been designed to help parents and teachers reinforce basic skills with their children. *Practice Makes Perfect* reviews basic math skills for children in grades 3 and 4. The math focus is on fractions. While it would be impossible to include all concepts taught in grades 3 and 4 in this book, the following basic objectives are reinforced through practice exercises. These objectives support math standards established on a district, state, or national level. (Refer to the Table of Contents for the specific objectives of each practice page.)

- identifying, writing, and ordering fractions
- naming and comparing proper fractions
- identifying proper fractions as a whole or part of a set
- using models to understand fractions
- writing, recognizing, and comparing equivalent fractions

- writing fractions in simplest form
- adding and subtracting fractions with like denominators
- writing mixed and improper fractions
- representing money as fractions

There are 36 practice pages organized sequentially, so children can build their knowledge from more basic skills to higher-level math skills. (**Note:** Have children show all work where computation is necessary to solve a problem. For multiple-choice responses on practice pages, children can fill in the letter choice or circle the answer.) Following the practice pages are six test practices. These provide children with multiple-choice test items to help prepare them for standardized tests administered in schools. As your child completes each test, he or she should fill in the correct bubbles on the answer sheet (page 46). To correct the test pages and the practice pages in this book, use the Test Practice answer key provided on pages 47 and 48.

How to Make the Most of This Book

Here are some useful ideas for optimizing the practice pages in this book:

- Set aside a specific place in your home to work on the practice pages. Keep it neat and tidy with materials on hand.
- Set up a certain time of day to work on the practice pages. This will establish consistency. An alternative is to look for times in your day or week that are less hectic and conducive to practicing skills.
- Keep all practice sessions with your child positive and constructive. If the mood becomes tense or you and your child are frustrated, set the book aside and look for another time to practice with your child.
- Help with instructions if necessary. If your child is having difficulty understanding what to do or how to get started, work through the first problem with him or her.
- Review the work your child has done. This serves as reinforcement and provides further practice.
- Allow your child to use whatever writing instruments he or she prefers. For example, colored pencils can add variety and pleasure to drill work.
- Pay attention to the areas in which your child has the most difficulty. Provide extra guidance and exercises in those areas. Allowing children to use drawings and manipulatives, such as coins, tiles, game markers, or flash cards, can help them grasp difficult concepts more easily.
- Look for ways to make real-life applications to the skills being reinforced.

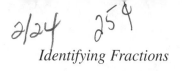
Practice 1 ᗧ ᗡ ᗧ ᗡ ᗧ ᗡ ᗧ ᗡ ᗧ ᗡ ᗧ ᗡ ᗧ ᗡ ᗡ ᗧ

A **fraction** is a number that names part of a whole thing. The number at the top is the numerator. It tells how many parts of the whole are present. The number at the bottom is the denominator. It tells how many parts there are in all.

Examples

$\frac{1}{2}$ (There are two parts in the circle. One part is gray. Therefore, the fraction is $\frac{1}{2}$.)

$\frac{3}{4}$ (There are four parts in the square. Three parts are gray. The fraction is $\frac{3}{4}$.)

Write a fraction for each picture.

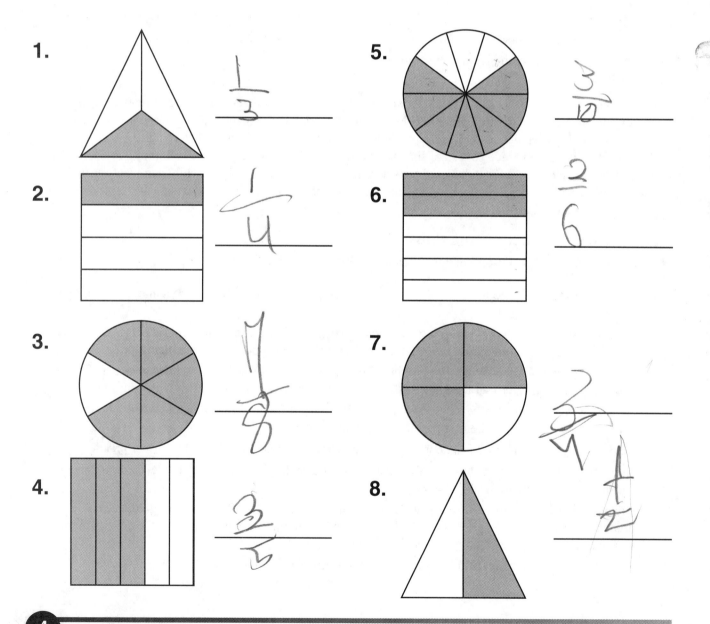

1. $\frac{1}{3}$

2. $\frac{1}{4}$

3. $\frac{4}{6}$

4. $\frac{3}{5}$

5. $\frac{3}{5}$

6. $\frac{2}{6}$

7. $\frac{3}{4}$

8. $\frac{1}{2}$

© *Teacher Created Resources, Inc.*

Practice 2

A *fraction* is a part (or parts) of a whole item or shape.

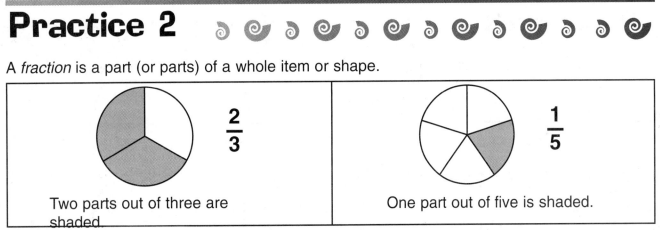

$\dfrac{2}{3}$

Two parts out of three are shaded.

$\dfrac{1}{5}$

One part out of five is shaded.

Directions: Look at each shape. Write the fraction that tells how many parts of the whole object are shaded. The first one has already been done for you.

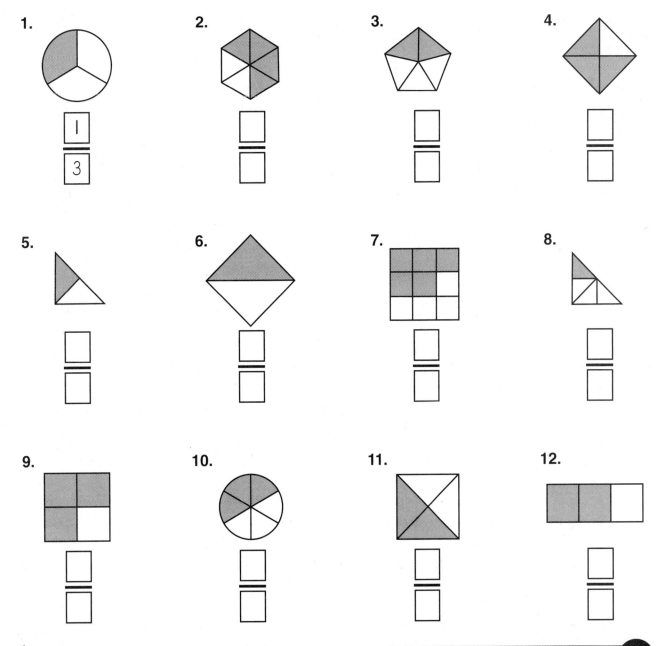

Practice 3 🌀 🐚 🌀 🐚 🌀 🐚 🌀 🐚 🌀 🐚 🌀 🐚

Directions: Color in each circle to show the correct fraction.

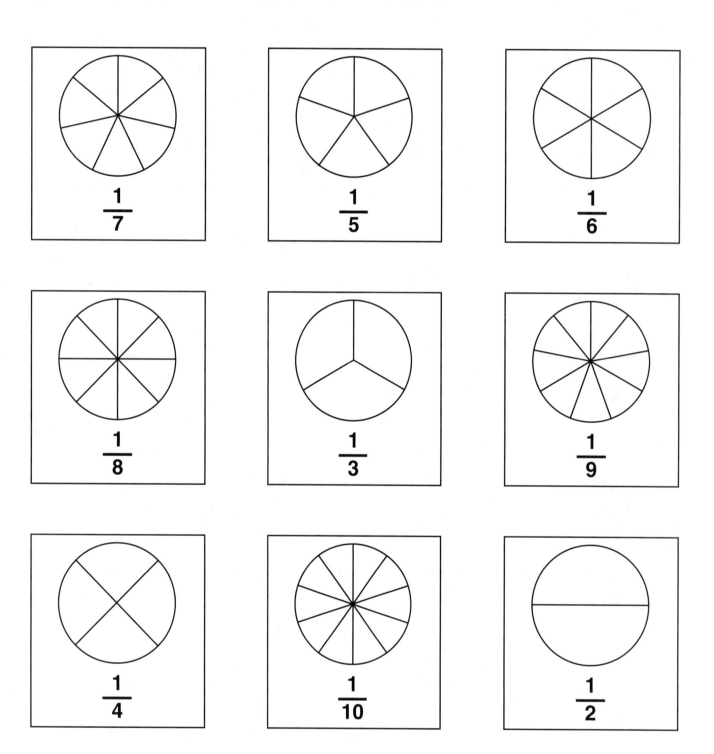

Directions: Write these fractions in order from largest to smallest.

_____, _____, _____, _____, _____, _____, _____, _____, _____

Practice 4

1. What fraction of the circle is *shaded*?

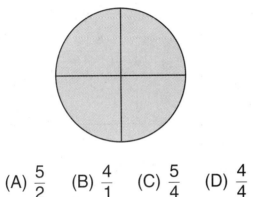

(A) $\frac{5}{2}$ (B) $\frac{4}{1}$ (C) $\frac{5}{4}$ (D) $\frac{4}{4}$

2. What fraction of the circle is *shaded*?

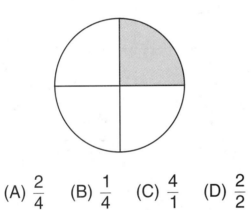

(A) $\frac{2}{4}$ (B) $\frac{1}{4}$ (C) $\frac{4}{1}$ (D) $\frac{2}{2}$

3. What fraction of the circle is *shaded*?

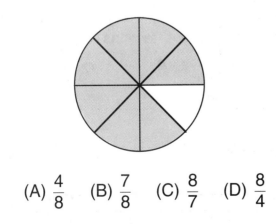

(A) $\frac{4}{8}$ (B) $\frac{7}{8}$ (C) $\frac{8}{7}$ (D) $\frac{8}{4}$

4. What fraction of the circle is *shaded*?

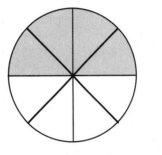

(A) $\frac{5}{4}$ (B) $\frac{4}{5}$ (C) $\frac{4}{8}$ (D) $\frac{8}{4}$

5. What fraction of the circle is *shaded*?

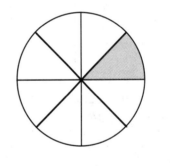

(A) $\frac{2}{4}$ (B) $\frac{4}{2}$ (C) $\frac{1}{8}$ (D) $\frac{8}{1}$

6. What fraction of the circle is *shaded*?

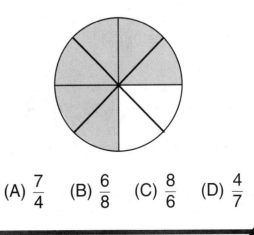

(A) $\frac{7}{4}$ (B) $\frac{6}{8}$ (C) $\frac{8}{6}$ (D) $\frac{4}{7}$

Practice 5

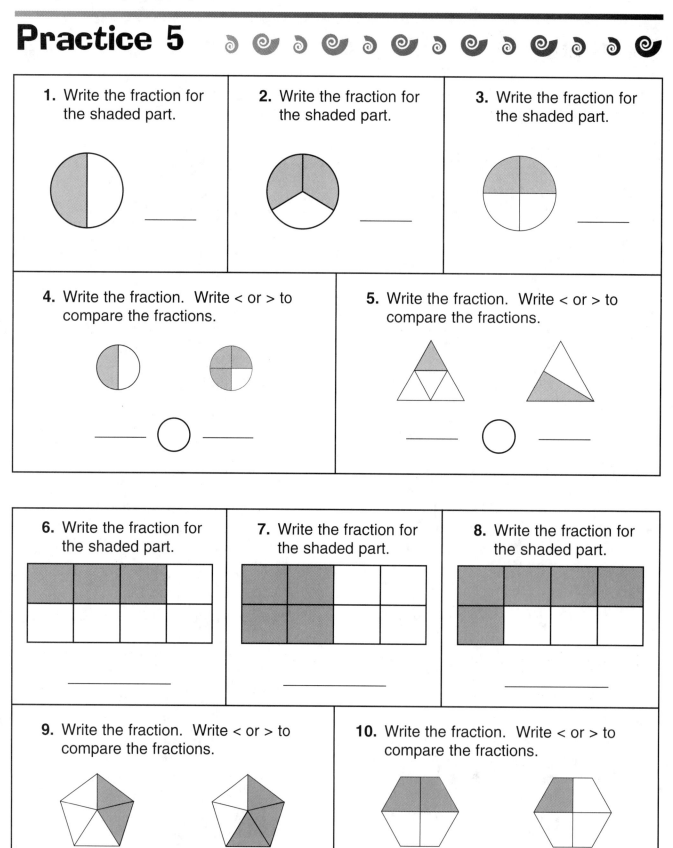

1. Write the fraction for the shaded part.

2. Write the fraction for the shaded part.

3. Write the fraction for the shaded part.

4. Write the fraction. Write < or > to compare the fractions.

5. Write the fraction. Write < or > to compare the fractions.

6. Write the fraction for the shaded part.

7. Write the fraction for the shaded part.

8. Write the fraction for the shaded part.

9. Write the fraction. Write < or > to compare the fractions.

10. Write the fraction. Write < or > to compare the fractions.

Practice 6

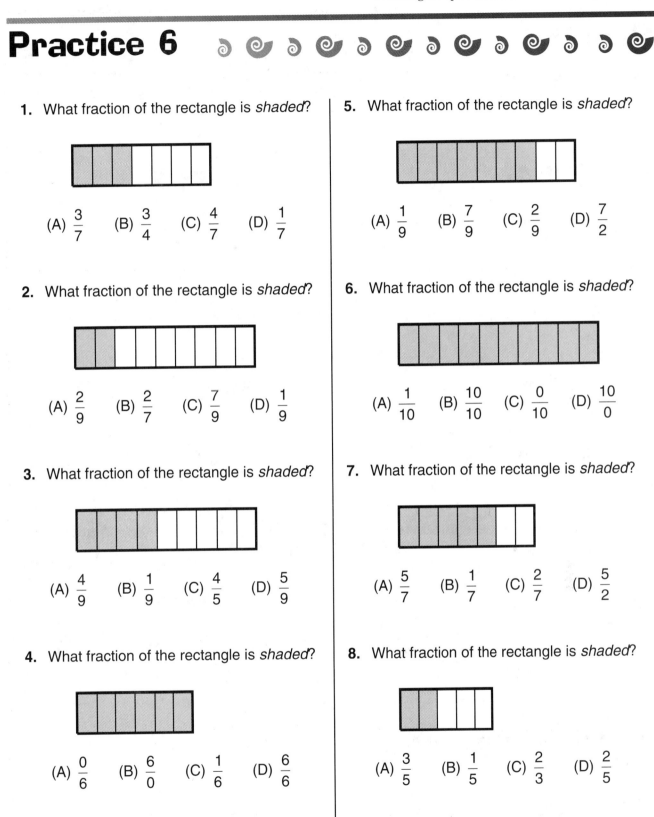

1. What fraction of the rectangle is *shaded*?

   (A) $\frac{3}{7}$ (B) $\frac{3}{4}$ (C) $\frac{4}{7}$ (D) $\frac{1}{7}$

2. What fraction of the rectangle is *shaded*?

   (A) $\frac{2}{9}$ (B) $\frac{2}{7}$ (C) $\frac{7}{9}$ (D) $\frac{1}{9}$

3. What fraction of the rectangle is *shaded*?

   (A) $\frac{4}{9}$ (B) $\frac{1}{9}$ (C) $\frac{4}{5}$ (D) $\frac{5}{9}$

4. What fraction of the rectangle is *shaded*?

   (A) $\frac{0}{6}$ (B) $\frac{6}{0}$ (C) $\frac{1}{6}$ (D) $\frac{6}{6}$

5. What fraction of the rectangle is *shaded*?

   (A) $\frac{1}{9}$ (B) $\frac{7}{9}$ (C) $\frac{2}{9}$ (D) $\frac{7}{2}$

6. What fraction of the rectangle is *shaded*?

   (A) $\frac{1}{10}$ (B) $\frac{10}{10}$ (C) $\frac{0}{10}$ (D) $\frac{10}{0}$

7. What fraction of the rectangle is *shaded*?

   (A) $\frac{5}{7}$ (B) $\frac{1}{7}$ (C) $\frac{2}{7}$ (D) $\frac{5}{2}$

8. What fraction of the rectangle is *shaded*?

   (A) $\frac{3}{5}$ (B) $\frac{1}{5}$ (C) $\frac{2}{3}$ (D) $\frac{2}{5}$

Practice 7

1. What fraction of the circles are white?

○●●●●○

(A) $\frac{4}{6}$ (B) $\frac{5}{2}$ (C) $\frac{2}{4}$ (D) $\frac{2}{6}$

2. What fraction of the circles are black?

○○●●○○○

(A) $\frac{5}{2}$ (B) $\frac{5}{7}$ (C) $\frac{2}{7}$ (D) $\frac{2}{5}$

3. What fraction of the circles are black?

○●●○○○○●○○

(A) $\frac{3}{10}$ (B) $\frac{7}{10}$ (C) $\frac{7}{3}$ (D) $\frac{3}{7}$

4. What fraction of the circles are white?

○○●

(A) $\frac{1}{3}$ (B) $\frac{1}{2}$ (C) $\frac{5}{3}$ (D) $\frac{2}{3}$

5. What fraction of the circles are white?

○○○●●○●○●○●

(A) $\frac{6}{11}$ (B) $\frac{5}{6}$ (C) $\frac{5}{11}$ (D) $\frac{6}{5}$

6. What fraction of the circles are white?

●●○○○○

(A) $\frac{2}{6}$ (B) $\frac{4}{6}$ (C) $\frac{2}{4}$ (D) $\frac{5}{3}$

7. What fraction of the circles are black?

●●●●●●●●○○○

(A) $\frac{8}{3}$ (B) $\frac{8}{11}$ (C) $\frac{3}{8}$ (D) $\frac{3}{11}$

8. What fraction of the circles are black?

○●○○○○

(A) $\frac{1}{6}$ (B) $\frac{5}{6}$ (C) $\frac{1}{5}$ (D) $\frac{5}{2}$

Practice 8

1. Write the fraction shown by the point on the number line.

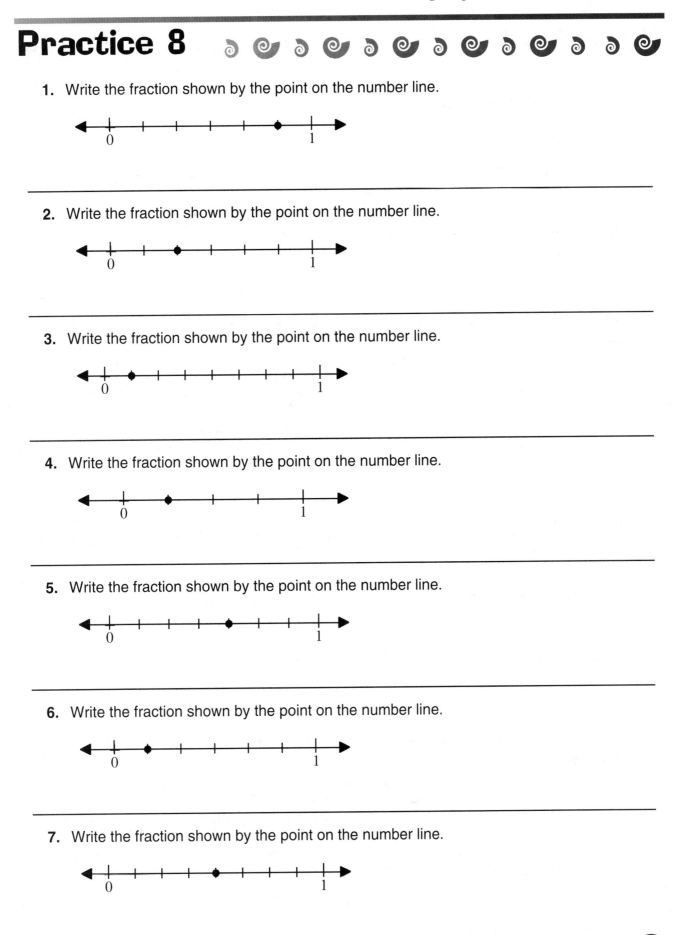

2. Write the fraction shown by the point on the number line.

3. Write the fraction shown by the point on the number line.

4. Write the fraction shown by the point on the number line.

5. Write the fraction shown by the point on the number line.

6. Write the fraction shown by the point on the number line.

7. Write the fraction shown by the point on the number line.

Practice 9

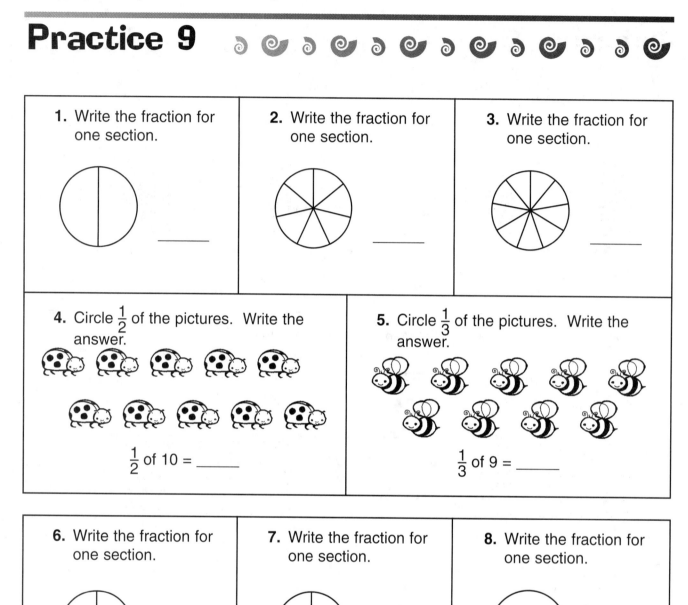

1. Write the fraction for one section.

2. Write the fraction for one section.

3. Write the fraction for one section.

4. Circle $\frac{1}{2}$ of the pictures. Write the answer.

$\frac{1}{2}$ of 10 = _____

5. Circle $\frac{1}{3}$ of the pictures. Write the answer.

$\frac{1}{3}$ of 9 = _____

6. Write the fraction for one section.

7. Write the fraction for one section.

8. Write the fraction for one section.

9. Divide the pictures into 3 equal sets. Complete the problem.

$\frac{2}{3}$ of 12 = _____

10. Divide the pictures into 6 equal sets. Complete the problem.

$\frac{3}{6}$ of 12 = _____

Practice 10

1. Name and compare the shaded fraction parts using >, <, or =.

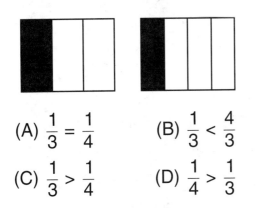

(A) $\frac{1}{3} = \frac{1}{4}$

(B) $\frac{1}{3} < \frac{4}{3}$

(C) $\frac{1}{3} > \frac{1}{4}$

(D) $\frac{1}{4} > \frac{1}{3}$

2. Name and compare the shaded fraction parts using >, <, or =.

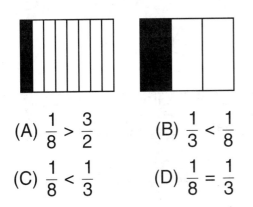

(A) $\frac{1}{8} > \frac{3}{2}$

(B) $\frac{1}{3} < \frac{1}{8}$

(C) $\frac{1}{8} < \frac{1}{3}$

(D) $\frac{1}{8} = \frac{1}{3}$

3. Name and compare the shaded fraction parts using >, <, or =.

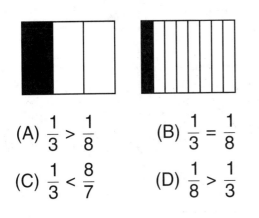

(A) $\frac{1}{3} > \frac{1}{8}$

(B) $\frac{1}{3} = \frac{1}{8}$

(C) $\frac{1}{3} < \frac{8}{7}$

(D) $\frac{1}{8} > \frac{1}{3}$

4. Name and compare the shaded fraction parts using >, <, or =.

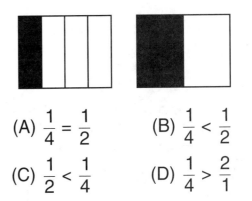

(A) $\frac{1}{4} = \frac{1}{2}$

(B) $\frac{1}{4} < \frac{1}{2}$

(C) $\frac{1}{2} < \frac{1}{4}$

(D) $\frac{1}{4} > \frac{2}{1}$

5. Name and compare the shaded fraction parts using >, <, or =.

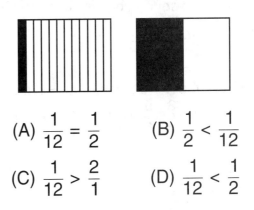

(A) $\frac{1}{12} = \frac{1}{2}$

(B) $\frac{1}{2} < \frac{1}{12}$

(C) $\frac{1}{12} > \frac{2}{1}$

(D) $\frac{1}{12} < \frac{1}{2}$

6. Name and compare the shaded fraction parts using >, <, or =.

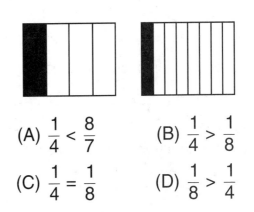

(A) $\frac{1}{4} < \frac{8}{7}$

(B) $\frac{1}{4} > \frac{1}{8}$

(C) $\frac{1}{4} = \frac{1}{8}$

(D) $\frac{1}{8} > \frac{1}{4}$

Practice 11

1. Order these fractions from *least* to *greatest*: $\frac{2}{7}, \frac{7}{7}, \frac{6}{7}, \frac{5}{7}$

2. Order these fractions from *least* to *greatest*: $\frac{11}{11}, \frac{3}{11}, \frac{5}{11}, \frac{1}{11}$

3. Name and compare the *shaded* fraction parts using >, <, or =.

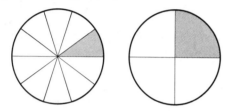

4. Order these fractions from *least* to *greatest*: $\frac{3}{13}, \frac{13}{13}, \frac{8}{13}, \frac{2}{13}$

5. Order these fractions from *least* to *greatest*: $\frac{3}{5}, \frac{5}{5}, \frac{2}{5}, \frac{1}{5}$

6. Name and compare the *shaded* fraction parts using >, <, or =.

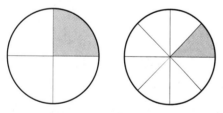

7. Order these fractions from *least* to *greatest*: $\frac{2}{11}, \frac{11}{11}, \frac{9}{11}, \frac{7}{11}$

8. Order these fractions from *least* to *greatest*: $\frac{2}{13}, \frac{8}{13}, \frac{13}{13}, \frac{12}{13}$

9. Name and compare the *shaded* fraction parts using >, <, or =.

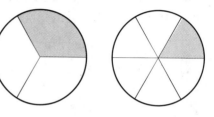

10. Order these fractions from *least* to *greatest*: $\frac{2}{5}, \frac{5}{5}, \frac{1}{5}, \frac{3}{5}$

11. Name and compare the *shaded* fraction parts using >, <, or =.

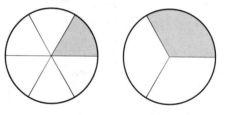

12. Order these fractions from *least* to *greatest*: $\frac{4}{5}, \frac{5}{5}, \frac{2}{5}, \frac{3}{5}$

13. Order these fractions from *least* to *greatest*: $\frac{5}{7}, \frac{3}{7}, \frac{7}{7}, \frac{4}{7}$

14. Order these fractions from *least* to *greatest*: $\frac{13}{13}, \frac{4}{13}, \frac{2}{13}, \frac{12}{13}$

15. Name and compare the *shaded* fraction parts using >, <, or =.

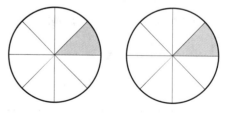

Practice 12

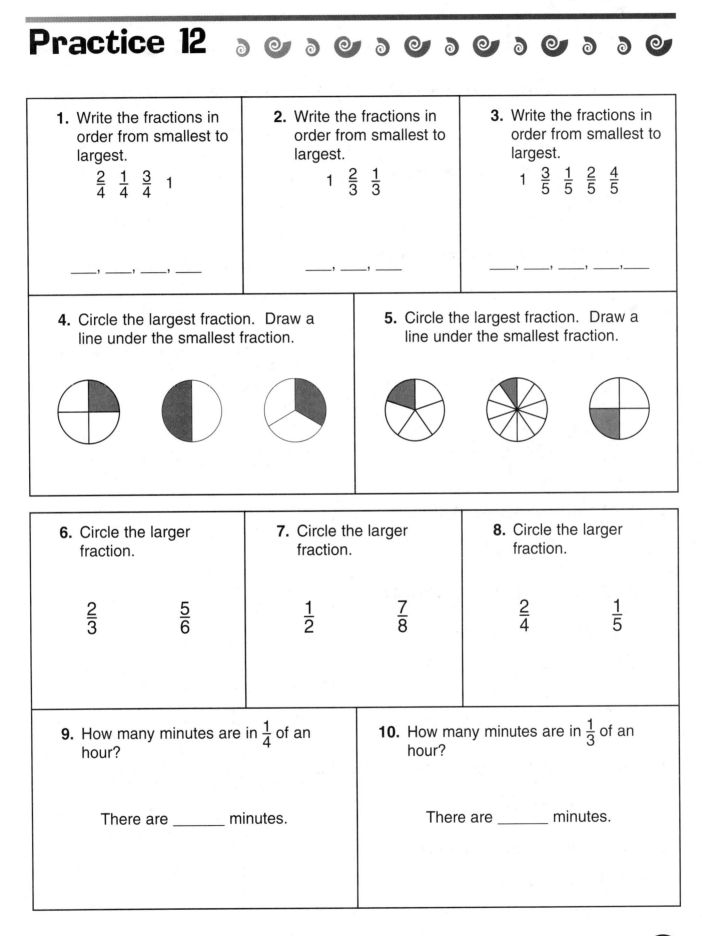

1. Write the fractions in order from smallest to largest.

$\frac{2}{4}$ $\frac{1}{4}$ $\frac{3}{4}$ 1

___, ___, ___, ___

2. Write the fractions in order from smallest to largest.

1 $\frac{2}{3}$ $\frac{1}{3}$

___, ___, ___

3. Write the fractions in order from smallest to largest.

1 $\frac{3}{5}$ $\frac{1}{5}$ $\frac{2}{5}$ $\frac{4}{5}$

___, ___, ___, ___,___

4. Circle the largest fraction. Draw a line under the smallest fraction.

5. Circle the largest fraction. Draw a line under the smallest fraction.

6. Circle the larger fraction.

$\frac{2}{3}$ $\frac{5}{6}$

7. Circle the larger fraction.

$\frac{1}{2}$ $\frac{7}{8}$

8. Circle the larger fraction.

$\frac{2}{4}$ $\frac{1}{5}$

9. How many minutes are in $\frac{1}{4}$ of an hour?

There are _____ minutes.

10. How many minutes are in $\frac{1}{3}$ of an hour?

There are _____ minutes.

Practice 13

1. What equivalent fractions are *shaded?*

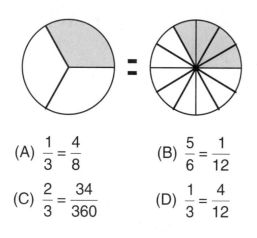

(A) $\dfrac{1}{3} = \dfrac{4}{8}$ (B) $\dfrac{5}{6} = \dfrac{1}{12}$

(C) $\dfrac{2}{3} = \dfrac{34}{360}$ (D) $\dfrac{1}{3} = \dfrac{4}{12}$

2. What equivalent fractions are *shaded?*

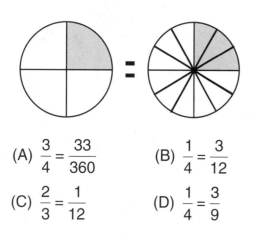

(A) $\dfrac{3}{4} = \dfrac{33}{360}$ (B) $\dfrac{1}{4} = \dfrac{3}{12}$

(C) $\dfrac{2}{3} = \dfrac{1}{12}$ (D) $\dfrac{1}{4} = \dfrac{3}{9}$

3. What equivalent fractions are *shaded?*

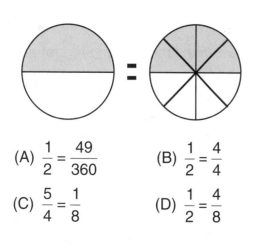

(A) $\dfrac{1}{2} = \dfrac{49}{360}$ (B) $\dfrac{1}{2} = \dfrac{4}{4}$

(C) $\dfrac{5}{4} = \dfrac{1}{8}$ (D) $\dfrac{1}{2} = \dfrac{4}{8}$

4. What equivalent fractions are *shaded?*

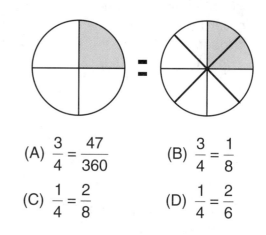

(A) $\dfrac{3}{4} = \dfrac{47}{360}$ (B) $\dfrac{3}{4} = \dfrac{1}{8}$

(C) $\dfrac{1}{4} = \dfrac{2}{8}$ (D) $\dfrac{1}{4} = \dfrac{2}{6}$

5. What equivalent fractions are *shaded?*

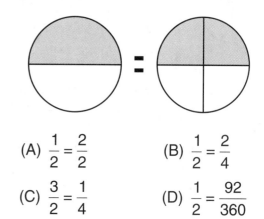

(A) $\dfrac{1}{2} = \dfrac{2}{2}$ (B) $\dfrac{1}{2} = \dfrac{2}{4}$

(C) $\dfrac{3}{2} = \dfrac{1}{4}$ (D) $\dfrac{1}{2} = \dfrac{92}{360}$

6. What equivalent fractions are *shaded?*

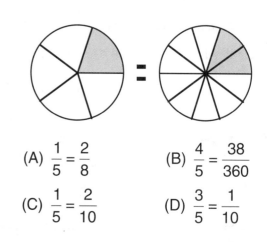

(A) $\dfrac{1}{5} = \dfrac{2}{8}$ (B) $\dfrac{4}{5} = \dfrac{38}{360}$

(C) $\dfrac{1}{5} = \dfrac{2}{10}$ (D) $\dfrac{3}{5} = \dfrac{1}{10}$

Practice 14

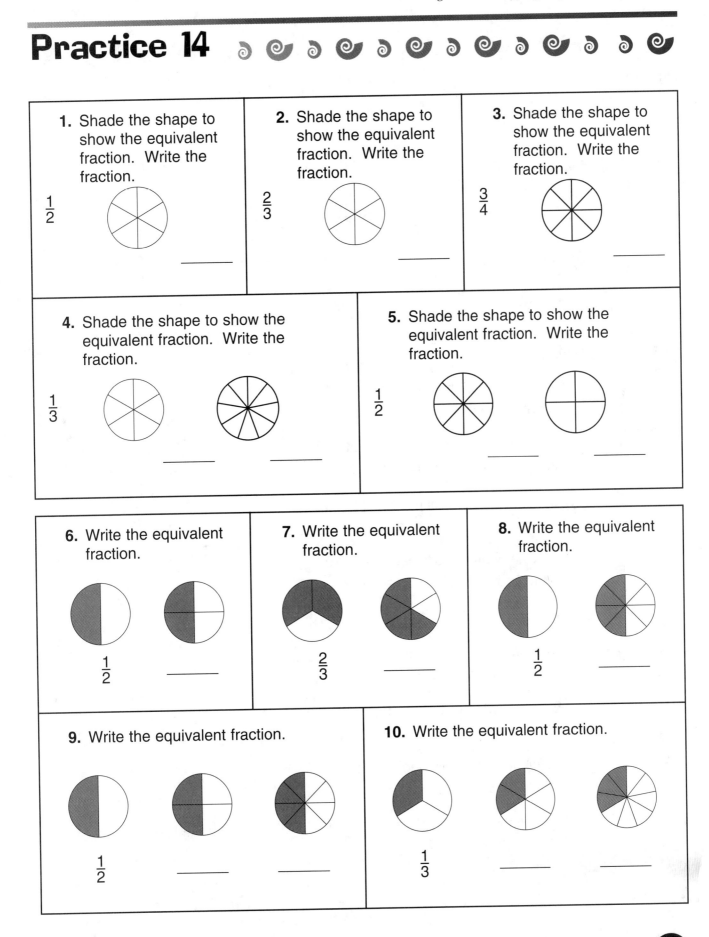

1. Shade the shape to show the equivalent fraction. Write the fraction.

$\frac{1}{2}$ _____

2. Shade the shape to show the equivalent fraction. Write the fraction.

$\frac{2}{3}$ _____

3. Shade the shape to show the equivalent fraction. Write the fraction.

$\frac{3}{4}$ _____

4. Shade the shape to show the equivalent fraction. Write the fraction.

$\frac{1}{3}$ _____

5. Shade the shape to show the equivalent fraction. Write the fraction.

$\frac{1}{2}$ _____ _____

6. Write the equivalent fraction.

$\frac{1}{2}$ _____

7. Write the equivalent fraction.

$\frac{2}{3}$ _____

8. Write the equivalent fraction.

$\frac{1}{2}$ _____

9. Write the equivalent fraction.

$\frac{1}{2}$ _____ _____

10. Write the equivalent fraction.

$\frac{1}{3}$ _____ _____

Practice 15 ⟳ ⟲ ⟳ ⟲ ⟳ ⟲ ⟳ ⟲ ⟳ ⟲ ⟳ ⟲

Directions: Circle the fraction in each group that is equivalent to $\frac{1}{2}$.

1.   **a.** $\frac{1}{4}$ **b.** $\frac{2}{4}$ **c.** $\frac{2}{3}$

2.   **a.** $\frac{7}{14}$ **b.** $\frac{2}{6}$ **c.** $\frac{3}{9}$

3.   **a.** $\frac{6}{9}$ **b.** $\frac{9}{18}$ **c.** $\frac{9}{12}$

4.   **a.** $\frac{1}{4}$ **b.** $\frac{12}{24}$ **c.** $\frac{12}{36}$

5.   **a.** $\frac{8}{14}$ **b.** $\frac{14}{26}$ **c.** $\frac{10}{20}$

Directions: Multiply.

6. $\frac{3}{7} = \frac{}{21}$ By what was 7 multiplied to get 21? _____
     What is the missing numerator? _____

7. $\frac{4}{9} = \frac{}{18}$ By what was 9 multiplied to get 18? _____
     What is the missing numerator? _____

$$\frac{75}{100} = \frac{3}{4}$$

8. $\frac{2}{3} = \frac{}{18}$ By what was 3 multiplied to get 18? _____
     What is the missing numerator? _____

9. $\frac{2}{5} = \frac{}{10}$ By what was 5 multiplied to get 10? _____
     What is the missing numerator? _____

$$\frac{60}{85} = \frac{12}{17}$$

Directions: Divide.

10. $\frac{9}{12} = \frac{}{4}$ By what was 12 divided to get 4? _____
      What is the missing numerator? _____

11. $\frac{8}{16} = \frac{}{2}$ By what was 16 divided to get 2? _____
      What is the missing numerator? _____

$$\frac{33}{99} = \frac{1}{3}$$

12. $\frac{10}{24} = \frac{}{12}$ By what was 24 divided to get 12? _____
      What is the missing numerator? _____

Directions: Which fraction (a, b, or c) is not equivalent to the given fraction?

13. $\frac{2}{3}$ **a.** $\frac{2}{6}$ **b.** $\frac{6}{9}$ **c.** $\frac{8}{12}$

14. $\frac{1}{5}$ **a.** $\frac{3}{15}$ **b.** $\frac{2}{10}$ **c.** $\frac{1}{10}$

#3325 Practice Makes Perfect: Fractions

Practice 16 ᕲ ᕮ ᕲ ᕮ ᕲ ᕮ ᕲ ᕮ ᕲ ᕮ ᕲ ᕲ ᕮ

Write each fraction in simplest form.

1. $\frac{6}{8} =$ _____

2. $\frac{4}{10} =$ _____

3. $\frac{3}{9} =$ _____

4. $\frac{5}{10} =$ _____

5. $\frac{3}{6} =$ _____

6. $\frac{2}{8} =$ _____

7. $\frac{4}{12} =$ _____

8. $\frac{3}{12} =$ _____

9. $\frac{5}{15} =$ _____

10. $\frac{2}{10} =$ _____

11. $\frac{8}{16} =$ _____

12. $\frac{2}{14} =$ _____

Write **yes** if the fraction is written in simplest form or **no** if it is not in simplest form.

13. $\frac{2}{3}$

14. $\frac{4}{6}$

15. $\frac{1}{5}$

16. $\frac{3}{4}$

_____

_____

_____

_____

17. $\frac{8}{10}$

18. $\frac{7}{9}$

19. $\frac{5}{8}$

20. $\frac{1}{10}$

_____

_____

_____

_____

21. $\frac{3}{14}$

22. $\frac{4}{16}$

23. $\frac{4}{8}$

24. $\frac{2}{4}$

_____

_____

_____

_____

Practice 17

1. Simplify the fraction.

$$\frac{6}{8} =$$

2. Simplify the fraction.

$$\frac{5}{10} =$$

3. Simplify the fraction.

$$\frac{4}{8} =$$

4. Solve.

A recipe calls for $\frac{3}{9}$ cup of butter. Kevin only has measuring cups for $\frac{1}{2}$, $\frac{1}{3}$, and $\frac{1}{4}$. Which cup should he use?

He should use the _____ measuring cup.

5. Solve.

Jolie needs to put $\frac{4}{6}$ gallon of gas in the lawn mower. She only has canisters that measure $\frac{1}{3}$, $\frac{2}{3}$, or $\frac{1}{4}$. Which canister should she use?

She should use the _____ canister.

6. Circle the equivalent fraction.

$$\frac{1}{2}$$

$$\frac{3}{4} \qquad \frac{3}{5} \qquad \frac{5}{10}$$

7. Circle the equivalent fraction.

$$\frac{1}{6}$$

$$\frac{2}{8} \qquad \frac{4}{8} \qquad \frac{2}{12}$$

8. Circle the equivalent fraction.

$$\frac{3}{4}$$

$$\frac{2}{3} \qquad \frac{6}{8} \qquad \frac{3}{9}$$

9. Write two equivalent fractions for each fraction.

$$\frac{2}{4} = \rule{1cm}{0.15mm} \text{ and } \rule{1cm}{0.15mm}$$

$$\frac{3}{9} = \rule{1cm}{0.15mm} \text{ and } \rule{1cm}{0.15mm}$$

10. Write two equivalent fractions for each fraction.

$$\frac{6}{12} = \rule{1cm}{0.15mm} \text{ and } \rule{1cm}{0.15mm}$$

$$\frac{4}{8} = \rule{1cm}{0.15mm} \text{ and } \rule{1cm}{0.15mm}$$

Practice 18

1. $\frac{2}{5} + \frac{1}{5} =$ (A) $\frac{2}{25}$ (B) $\frac{3}{10}$ (C) $\frac{5}{3}$ (D) $\frac{3}{5}$

2. $\frac{3}{11} + \frac{1}{11} =$ (A) $\frac{4}{22}$ (B) $\frac{4}{11}$ (C) $\frac{11}{4}$ (D) $\frac{3}{121}$

3. $\frac{6}{11} + \frac{1}{11} =$ (A) $\frac{7}{11}$ (B) $\frac{6}{121}$ (C) $\frac{11}{7}$ (D) $\frac{7}{22}$

4. $\frac{2}{11} + \frac{1}{11} =$ (A) $\frac{11}{3}$ (B) $\frac{3}{11}$ (C) $\frac{3}{22}$ (D) $\frac{2}{121}$

5. $\frac{1}{7} + \frac{1}{7} =$ (A) $\frac{1}{49}$ (B) $\frac{2}{14}$ (C) $\frac{2}{7}$ (D) $\frac{7}{2}$

6. $\frac{1}{11} + \frac{1}{11} =$ (A) $\frac{2}{22}$ (B) $\frac{2}{11}$ (C) $\frac{1}{121}$ (D) $\frac{11}{2}$

7. $\frac{5}{7} + \frac{1}{7} =$ (A) $\frac{6}{7}$ (B) $\frac{5}{49}$ (C) $\frac{7}{6}$ (D) $\frac{6}{14}$

8. $\frac{9}{11} + \frac{1}{11} =$ (A) $\frac{11}{10}$ (B) $\frac{10}{11}$ (C) $\frac{9}{121}$ (D) $\frac{10}{22}$

9. $\frac{8}{11} + \frac{1}{11} =$ (A) $\frac{8}{121}$ (B) $\frac{9}{22}$ (C) $\frac{11}{9}$ (D) $\frac{9}{11}$

10. $\frac{1}{5} + \frac{1}{5} =$ (A) $\frac{2}{10}$ (B) $\frac{5}{2}$ (C) $\frac{2}{5}$ (D) $\frac{1}{25}$

Practice 19 ౭ ☙ ౭ ☙ ౭ ☙ ౭ ☙ ౭ ☙ ౭ ☙ ౭ ౭ ☙

Add the fractions and write your answer in simplest form where possible.

1. $\dfrac{2}{3}$
    $+ \dfrac{1}{3}$

2. $\dfrac{4}{15}$
    $+ \dfrac{2}{15}$

3. $\dfrac{1}{8}$
    $+ \dfrac{2}{8}$

4. $\dfrac{5}{9}$
    $+ \dfrac{2}{9}$

5. $\dfrac{3}{20}$
    $+ \dfrac{4}{20}$

6. $\dfrac{2}{7}$
    $+ \dfrac{1}{7}$

7. $\dfrac{3}{10}$
    $+ \dfrac{5}{10}$

8. $\dfrac{4}{5}$
    $+ \dfrac{1}{5}$

9. $\dfrac{7}{24}$
    $+ \dfrac{10}{24}$

10. $\dfrac{4}{7}$
    $+ \dfrac{1}{7}$

11. $\dfrac{3}{5}$
    $+ \dfrac{1}{5}$

12. $\dfrac{5}{8}$
    $+ \dfrac{2}{8}$

13. $\dfrac{6}{10}$
    $+ \dfrac{1}{10}$

14. $\dfrac{1}{3}$
    $+ \dfrac{1}{3}$

15. $\dfrac{10}{13}$
    $+ \dfrac{1}{13}$

16. $\dfrac{5}{12}$
    $+ \dfrac{3}{12}$

Practice 20 ❧ ❧ ❧ ❧ ❧ ❧ ❧ ❧ ❧ ❧ ❧ ❧ ❧ ❧

Add the fractions and write the answer in simplest form where possible.

1.

$\dfrac{4}{7} + \dfrac{2}{7} = $ _____

2.

$\dfrac{3}{5} + \dfrac{1}{5} = $ _____

3.

$\dfrac{5}{12} + \dfrac{2}{12} = $ _____

4.

$\dfrac{6}{8}$
$+ \dfrac{1}{8}$

5.

$\dfrac{10}{18}$
$+ \dfrac{4}{18}$

6.

$\dfrac{6}{14} + \dfrac{5}{14} = $ _____

7.

$\dfrac{8}{15}$
$+ \dfrac{1}{15}$

8.

$\dfrac{9}{17}$
$+ \dfrac{2}{17}$

9.

$\dfrac{4}{15}$
$+ \dfrac{1}{15}$

10.

$\dfrac{6}{13} + \dfrac{1}{13} = $ _____

11.

$\dfrac{5}{8}$
$+ \dfrac{2}{8}$

12.

$\dfrac{12}{19}$
$+ \dfrac{3}{19}$

13.

$\dfrac{7}{20}$
$+ \dfrac{8}{20}$

14.

$\dfrac{5}{16} + \dfrac{8}{16} = $ _____

15.

$\dfrac{4}{5}$
$+ \dfrac{1}{5}$

16.

$\dfrac{18}{30}$
$+ \dfrac{3}{30}$

Practice 21 ꩜ ꩜ ꩜ ꩜ ꩜ ꩜ ꩜ ꩜ ꩜ ꩜ ꩜ ꩜ ꩜ ꩜

Add the fractions and write your answer in simplest form where possible.

1. $1\frac{1}{2}$
   $+\ 1\frac{1}{2}$

2. $2\frac{1}{3}$
   $+\ 1\frac{1}{3}$

3. $5\frac{1}{6}$
   $+\ 2\frac{2}{6}$

4. $1\frac{1}{8}$
   $+\ \ \frac{3}{8}$

5. $6\frac{1}{9}$
   $+\ \ \frac{2}{9}$

6. $3\frac{2}{3}$
   $+\ 1\frac{1}{3}$

7. $5\frac{1}{4}$
   $+\ \ \frac{2}{4}$

8. $9\frac{1}{2}$
   $+\ 1\frac{1}{2}$

9. $2\frac{1}{6}$
   $+\ 1\frac{2}{6}$

10. $4\frac{2}{3}$
    $+\ 3\frac{1}{3}$

11. $2\frac{1}{8}$
    $+\ 1\frac{2}{8}$

12. $7\frac{1}{7}$
    $+\ 1\frac{3}{7}$

13. $2\frac{2}{8}$
    $+\ 3\frac{1}{8}$

14. $13\frac{1}{3}$
    $+\ 1\frac{2}{3}$

15. $22\frac{4}{5}$
    $+\ \ \frac{1}{5}$

16. $101\frac{1}{6}$
    $+\ 33\frac{2}{6}$

Practice 22

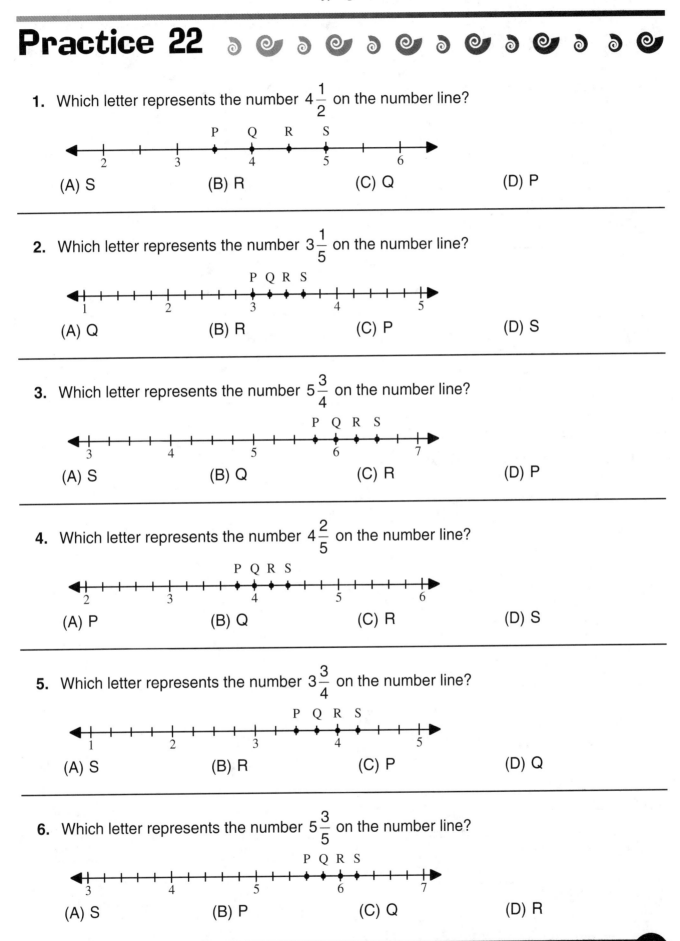

1. Which letter represents the number $4\frac{1}{2}$ on the number line?

   (A) S   (B) R   (C) Q   (D) P

2. Which letter represents the number $3\frac{1}{5}$ on the number line?

   (A) Q   (B) R   (C) P   (D) S

3. Which letter represents the number $5\frac{3}{4}$ on the number line?

   (A) S   (B) Q   (C) R   (D) P

4. Which letter represents the number $4\frac{2}{5}$ on the number line?

   (A) P   (B) Q   (C) R   (D) S

5. Which letter represents the number $3\frac{3}{4}$ on the number line?

   (A) S   (B) R   (C) P   (D) Q

6. Which letter represents the number $5\frac{3}{5}$ on the number line?

   (A) S   (B) P   (C) Q   (D) R

Practice 23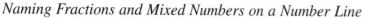

1. Name the fraction or the mixed number marked by the arrow.

2. Name the fraction or the mixed number marked by the arrow.

3. Name the fraction or the mixed number marked by the arrow.

4. Name the fraction or the mixed number marked by the arrow.

5. Name the fraction or the mixed number marked by the arrow.

6. Name the fraction or the mixed number marked by the arrow.

7. Name the fraction or the mixed number marked by the arrow.

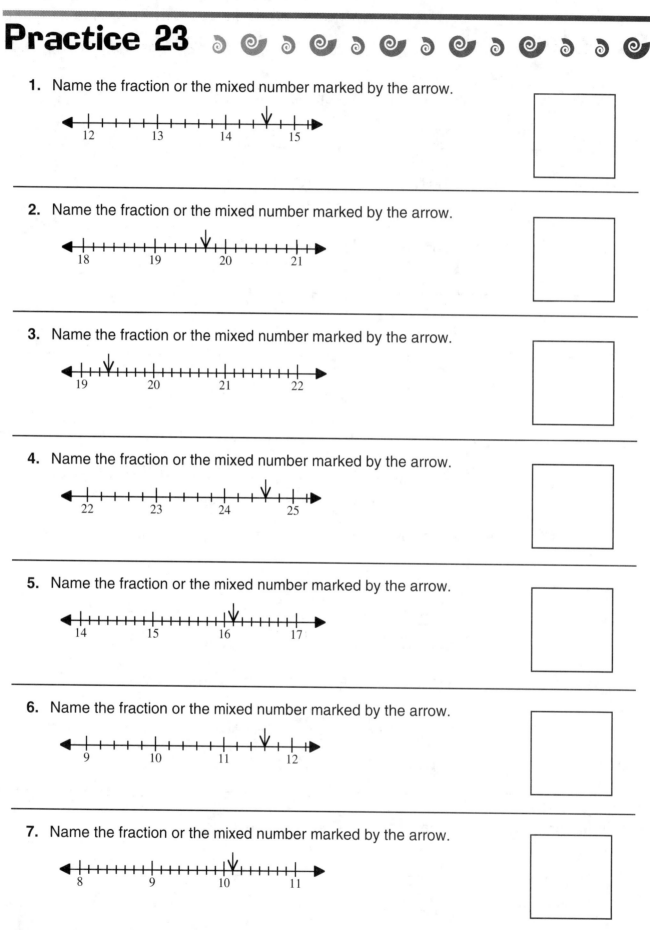

Practice 24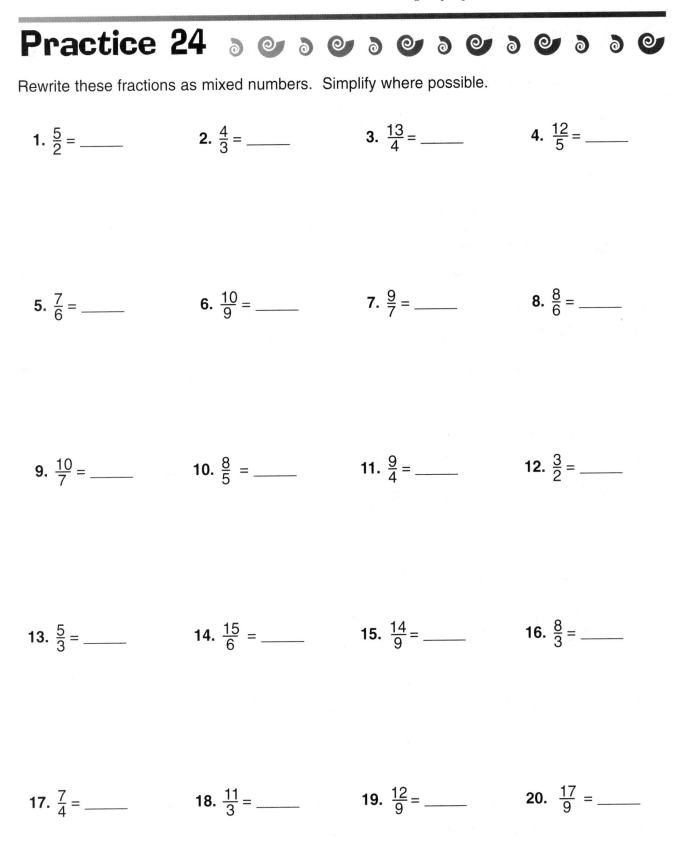

Rewrite these fractions as mixed numbers. Simplify where possible.

1. $\dfrac{5}{2}$ = _____

2. $\dfrac{4}{3}$ = _____

3. $\dfrac{13}{4}$ = _____

4. $\dfrac{12}{5}$ = _____

5. $\dfrac{7}{6}$ = _____

6. $\dfrac{10}{9}$ = _____

7. $\dfrac{9}{7}$ = _____

8. $\dfrac{8}{6}$ = _____

9. $\dfrac{10}{7}$ = _____

10. $\dfrac{8}{5}$ = _____

11. $\dfrac{9}{4}$ = _____

12. $\dfrac{3}{2}$ = _____

13. $\dfrac{5}{3}$ = _____

14. $\dfrac{15}{6}$ = _____

15. $\dfrac{14}{9}$ = _____

16. $\dfrac{8}{3}$ = _____

17. $\dfrac{7}{4}$ = _____

18. $\dfrac{11}{3}$ = _____

19. $\dfrac{12}{9}$ = _____

20. $\dfrac{17}{9}$ = _____

Practice 25 ༄ ༀ ༄ ༀ ༄ ༀ ༄ ༀ ༄ ༀ ༄ ༀ ༄ ༄ ༀ

1. $\dfrac{5}{7} - \dfrac{1}{7} =$ (A) 4   (B) $\dfrac{4}{7}$ (C) $\dfrac{6}{7}$ (D) $\dfrac{5}{7}$

2. $\dfrac{3}{17} - \dfrac{2}{17} =$ (A) 1   (B) $\dfrac{2}{17}$ (C) $\dfrac{1}{17}$ (D) $\dfrac{5}{17}$

3. $\dfrac{12}{13} - \dfrac{11}{13} =$ (A) $\dfrac{1}{13}$ (B) $\dfrac{2}{13}$ (C) $\dfrac{23}{13}$ (D) 1

4. $\dfrac{5}{13} - \dfrac{2}{13} =$ (A) 3   (B) $\dfrac{4}{13}$ (C) $\dfrac{7}{13}$ (D) $\dfrac{3}{13}$

5. $\dfrac{8}{13} - \dfrac{7}{13} =$ (A) $\dfrac{2}{13}$ (B) $\dfrac{15}{13}$ (C) $\dfrac{1}{13}$ (D) 1

6. $\dfrac{6}{11} - \dfrac{1}{11} =$ (A) $\dfrac{7}{11}$ (B) 5   (C) $\dfrac{6}{11}$ (D) $\dfrac{5}{11}$

7. At lunch, Harriet had $\dfrac{5}{8}$ cup of popcorn. She then gave $\dfrac{1}{8}$ cup of popcorn to her friend. How many cups of popcorn does she have left?

     (A) $\dfrac{4}{8}$ cup   (B) $\dfrac{5}{8}$ cup   (C) $\dfrac{16}{6}$ cups   (D) $\dfrac{4}{16}$ cup

8. Juanita has $\dfrac{3}{8}$ cup of raisins. She needs $\dfrac{1}{8}$ cup of raisins for her cookie recipe. How many cups of raisins will be left after she makes her cookies?

     (A) $\dfrac{2}{16}$ cup   (B) $\dfrac{3}{8}$ cup   (C) $\dfrac{2}{8}$ cup   (D) $\dfrac{16}{4}$ cups

Practice 26 ಲ ಲ ಲ ಲ ಲ ಲ ಲ ಲ ಲ ಲ ಲ ಲ ಲ ಲ

Subtract the fractions and write your answer in simplest form where possible.

1. $\dfrac{4}{12}$
    $-\dfrac{3}{12}$

2. $\dfrac{7}{8}$
    $-\dfrac{2}{8}$

3. $\dfrac{13}{16}$
    $-\dfrac{7}{16}$

4. $\dfrac{3}{6}$
    $-\dfrac{1}{6}$

5. $\dfrac{5}{6}$
    $-\dfrac{1}{6}$

6. $\dfrac{2}{3}$
    $-\dfrac{2}{3}$

7. $\dfrac{9}{10}$
    $-\dfrac{1}{10}$

8. $\dfrac{7}{8}$
    $-\dfrac{4}{8}$

9. $\dfrac{3}{4}$
    $-\dfrac{1}{4}$

10. $\dfrac{2}{5}$
    $-\dfrac{1}{5}$

11. $\dfrac{10}{11}$
    $-\dfrac{2}{11}$

12. $\dfrac{9}{10}$
    $-\dfrac{3}{10}$

13. $\dfrac{9}{10}$
    $-\dfrac{8}{10}$

14. $\dfrac{11}{16}$
    $-\dfrac{7}{16}$

15. $\dfrac{7}{12}$
    $-\dfrac{3}{12}$

16. $\dfrac{13}{14}$
    $-\dfrac{12}{14}$

Practice 27 ⟳ ⟳ ⟳ ⟳ ⟳ ⟳ ⟳ ⟳ ⟳ ⟳ ⟳ ⟳ ⟳ ⟳

Subtract the fractions and write your answers in simplest form where possible.

1. $\dfrac{3}{5}$
$-\dfrac{1}{5}$

2. $\dfrac{10}{13}$
$-\dfrac{8}{13}$

3. $\dfrac{12}{13}$
$-\dfrac{10}{13}$

4. $\dfrac{5}{12}$
$-\dfrac{1}{12}$

5. $\dfrac{7}{8}$
$-\dfrac{3}{8}$

6. $\dfrac{9}{10}$
$-\dfrac{2}{10}$

7. $\dfrac{8}{9}$
$-\dfrac{1}{9}$

8. $\dfrac{5}{7}$
$-\dfrac{2}{7}$

9. $\dfrac{5}{6}$
$-\dfrac{2}{6}$

10. $\dfrac{7}{11}$
$-\dfrac{4}{11}$

11. $\dfrac{15}{16}$
$-\dfrac{12}{16}$

12. $\dfrac{4}{7}$
$-\dfrac{1}{7}$

13. $\dfrac{9}{10}$
$-\dfrac{8}{10}$

14. $\dfrac{7}{9}$
$-\dfrac{7}{9}$

15. $\dfrac{11}{17}$
$-\dfrac{4}{17}$

16. $\dfrac{15}{19}$
$-\dfrac{12}{19}$

Practice 28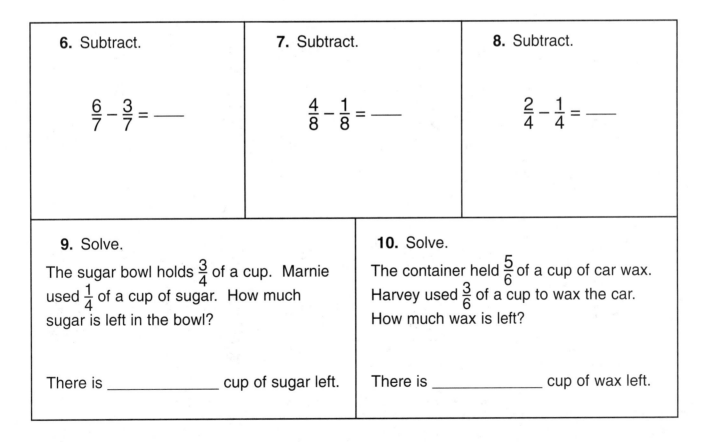

1. Subtract. $$\frac{6}{8} - \frac{5}{8} =$$	**2.** Subtract. $$\frac{4}{6} - \frac{3}{6} =$$	**3.** Add. $$\frac{4}{7} + \frac{1}{7} =$$

4. Rewrite the answer as a mixed fraction. $$\frac{2}{3} + \frac{2}{3} =$$	**5.** Rewrite the answer as a mixed fraction. $$\frac{2}{4} + \frac{3}{4} =$$

6. Subtract. $$\frac{6}{7} - \frac{3}{7} = \underline{\quad}$$	**7.** Subtract. $$\frac{4}{8} - \frac{1}{8} = \underline{\quad}$$	**8.** Subtract. $$\frac{2}{4} - \frac{1}{4} = \underline{\quad}$$

9. Solve. The sugar bowl holds $\frac{3}{4}$ of a cup. Marnie used $\frac{1}{4}$ of a cup of sugar. How much sugar is left in the bowl? There is _____ cup of sugar left.	**10.** Solve. The container held $\frac{5}{6}$ of a cup of car wax. Harvey used $\frac{3}{6}$ of a cup to wax the car. How much wax is left? There is _____ cup of wax left.

Practice 29 ⋑ ⋐ ⋑ ⋐ ⋑ ⋐ ⋑ ⋐ ⋑ ⋐ ⋑ ⋐ ⋑ ⋐ ⋑ ⋐

Subtract the fractions and write the answers in simplest form where possible.

1. $3\frac{3}{4}$
    $-1\frac{1}{4}$

2. $7\frac{7}{8}$
    $-\frac{2}{8}$

3. $10\frac{5}{6}$
    $-1\frac{3}{6}$

4. $6\frac{8}{9}$
    $-2\frac{1}{9}$

5. $8\frac{3}{4}$
    $-7\frac{1}{4}$

6. $15\frac{9}{10}$
    $-3\frac{2}{10}$

7. $19\frac{3}{5}$
    $-17\frac{1}{5}$

8. $8\frac{11}{12}$
    $-4\frac{2}{12}$

9. $5\frac{3}{7}$
    $-\frac{2}{7}$

10. $25\frac{14}{17}$
    $-12\frac{8}{17}$

11. $6\frac{5}{6}$
    $-2\frac{3}{6}$

12. $9\frac{16}{21}$
    $-2\frac{5}{21}$

13. $5\frac{9}{13}$
    $-3\frac{2}{13}$

14. $22\frac{5}{7}$
    $-9\frac{1}{7}$

15. $15\frac{7}{12}$
    $-7\frac{6}{12}$

16. $32\frac{8}{11}$
    $-4\frac{6}{11}$

Practice 30

Money can be written as a fraction.

1¢	**is the same as**	$\dfrac{1}{100}$

It takes 100 cents to make a dollar. One penny is one-hundredth of a dollar.

$0.41	**is the same as**	$\dfrac{41}{100}$

It takes 100 cents to make a dollar. 41¢ is 41 hundredths of a dollar.

Directions: Write each amount of money as a fraction. The first one has already been done for you.

1.
$$9¢ = \dfrac{9}{100}$$

2.
$$44¢ = \underline{\quad}$$

3.
$$98¢ = \underline{\quad}$$

4.
$$65¢ = \underline{\quad}$$

5.
$$39¢ = \underline{\quad}$$

6.
$$27¢ = \underline{\quad}$$

7.
$$\$0.73 = \underline{\quad}$$

8.
$$\$0.88 = \underline{\quad}$$

9.
$$\$0.23 = \underline{\quad}$$

10.
$$\$0.15 = \underline{\quad}$$

11.
$$\$0.50 = \underline{\quad}$$

12.
$$\$0.11 = \underline{\quad}$$

Directions: Use the > (greater than), < (less than), or = (equal to) symbols to compare the numbers. The first one has already been done for you.

13.
$$46¢ \enspace \boxed{<} \enspace \dfrac{99}{100}$$

14.
$$57¢ \enspace \bigcirc \enspace \dfrac{83}{100}$$

15.
$$18¢ \enspace \bigcirc \enspace \dfrac{18}{100}$$

16.
$$\$0.25 \enspace \bigcirc \enspace \dfrac{20}{100}$$

17.
$$\$0.63 \enspace \bigcirc \enspace \dfrac{63}{100}$$

18.
$$\$0.74 \enspace \bigcirc \enspace \dfrac{31}{100}$$

Practice 31 ꙮ ꙮ ꙮ ꙮ ꙮ ꙮ ꙮ ꙮ ꙮ ꙮ ꙮ ꙮ ꙮ ꙮ

Dollars and cents can be written as fractions too.

\$1.00 is the same as $\dfrac{100}{100}$ or $\dfrac{1}{1}$ or 1	**\$1.23** is the same as $\dfrac{123}{100}$ or $1\dfrac{23}{100}$

Directions: Write each amount of money as a mixed number. The first one has already been done for you.

1. $\$5.81 \; = \; 5\dfrac{81}{100}$

2. $\$2.71 \; = \; ---$

3. $\$3.19 \; = \; ---$

4. $\$1.86 \; = \; ---$

5. $\$4.04 \; = \; ---$

6. $\$5.11 \; = \; ---$

7. $\$1.63 \; = \; ---$

8. $\$2.10 \; = \; ---$

9. $\$3.07 \; = \; ---$

10. $\$4.29 \; = \; ---$

11. $\$7.99 \; = \; ---$

12. $\$3.00 \; = \; ---$

Directions: Convert each fraction to money. The first one has already been done for you.

13. $6\dfrac{19}{100} \; = \; \$6.19$

14. $2\dfrac{1}{100} \; =$

15. $4\dfrac{98}{100} \; =$

16. $3\dfrac{68}{100} \; =$

17. $1\dfrac{50}{100} \; =$

18. $2\dfrac{27}{100} \; =$

Practice 32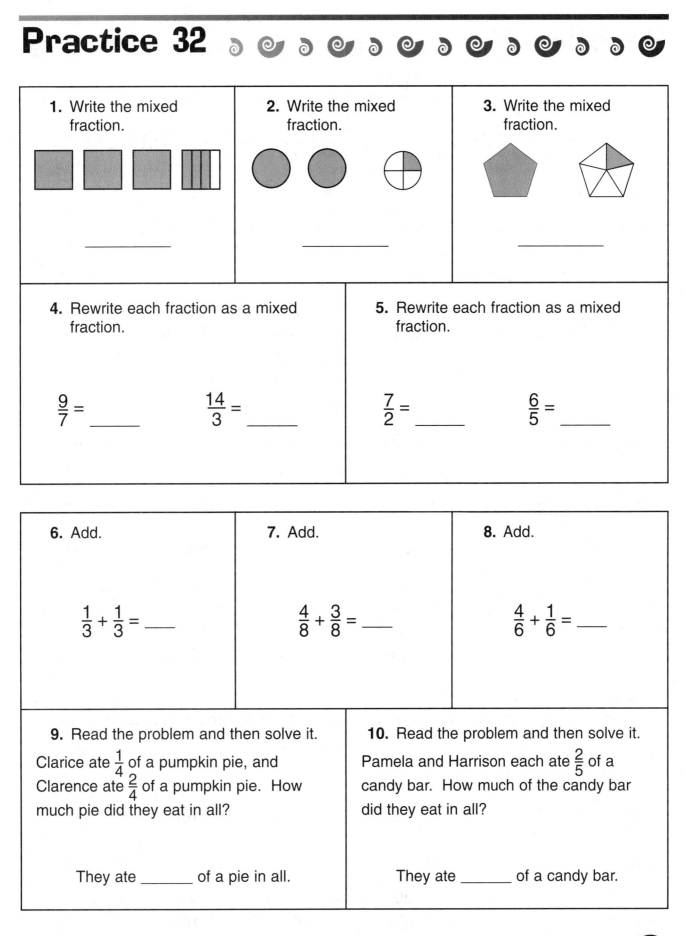

1. Write the mixed fraction.

_____

2. Write the mixed fraction.

_____

3. Write the mixed fraction.

_____

4. Rewrite each fraction as a mixed fraction.

$\dfrac{9}{7}$ = _____ $\dfrac{14}{3}$ = _____

5. Rewrite each fraction as a mixed fraction.

$\dfrac{7}{2}$ = _____ $\dfrac{6}{5}$ = _____

6. Add.

$\dfrac{1}{3} + \dfrac{1}{3} =$ ___

7. Add.

$\dfrac{4}{8} + \dfrac{3}{8} =$ ___

8. Add.

$\dfrac{4}{6} + \dfrac{1}{6} =$ ___

9. Read the problem and then solve it.

Clarice ate $\dfrac{1}{4}$ of a pumpkin pie, and Clarence ate $\dfrac{2}{4}$ of a pumpkin pie. How much pie did they eat in all?

They ate _____ of a pie in all.

10. Read the problem and then solve it.

Pamela and Harrison each ate $\dfrac{2}{5}$ of a candy bar. How much of the candy bar did they eat in all?

They ate _____ of a candy bar.

Practice 33

---

1. Subtract.

$$\frac{9}{10} - \frac{9}{10} = \underline{\quad}$$

2. Subtract.

$$\frac{8}{9} - \frac{4}{9} = \underline{\quad}$$

3. Subtract.

$$\frac{6}{7} - \frac{3}{7} = \underline{\quad}$$

---

4. Solve.

Ben was given $\frac{5}{8}$ of a pie. He ate $\frac{2}{8}$ of the pie. How much pie was left?

There was _____ pie left.

5. Solve.

Julia had $\frac{17}{24}$ of the crayons. Her dog ate $\frac{6}{24}$ of the crayons. How many crayons does Julia have left?

Julia has _____ of the crayons left.

---

6. How many turtles are in the fraction?

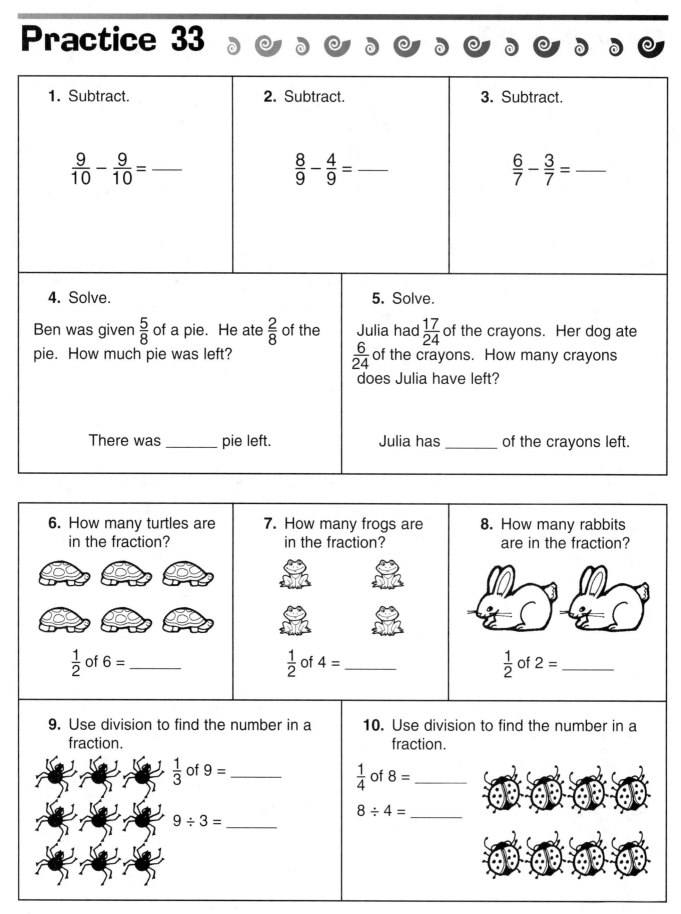

$\frac{1}{2}$ of 6 = _____

7. How many frogs are in the fraction?

$\frac{1}{2}$ of 4 = _____

8. How many rabbits are in the fraction?

$\frac{1}{2}$ of 2 = _____

---

9. Use division to find the number in a fraction.

$\frac{1}{3}$ of 9 = _____

$9 \div 3 =$ _____

10. Use division to find the number in a fraction.

$\frac{1}{4}$ of 8 = _____

$8 \div 4 =$ _____

Practice 34

---

1. Use division to find the number in a fraction.

$\frac{1}{2}$ of 10

_____

2. Use division to find the number in a fraction.

$\frac{1}{3}$ of 12

_____

3. Use division to find the number in a fraction.

$\frac{1}{4}$ of 16

_____

4. Use division to find the number in a fraction.

$\frac{1}{5}$ of 10 _____

$\frac{1}{4}$ of 20 _____

5. Use division to find the number in a fraction.

$\frac{1}{3}$ of 15 _____

$\frac{1}{2}$ of 24 _____

---

6. Write the fraction. Ana landed on red 3 out of 10 spins.

_____

7. Write the fraction. Deanna landed on blue 4 out of 5 spins.

_____

8. Write the fraction. Fred landed on red 2 out of 3 spins.

_____

9. Draw a spinner with 5 equal sections. Use red and blue to color the spinner. Make the spinner more likely to land on blue than red.

10. Draw a spinner with 5 equal sections. Use red and blue to color the spinner. Make the spinner more likely to land on red than blue.

Practice 35

In a **circle graph**, all the parts must add up to be a whole. Think of the parts like pieces that add up to one whole pie. Look at these pies and how they are divided into pieces.

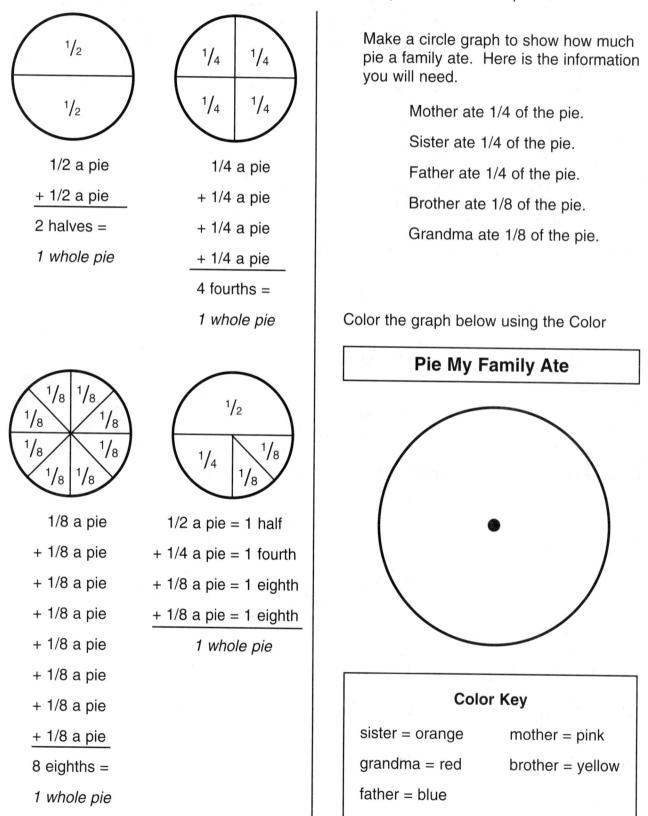

1/2 a pie

+ 1/2 a pie

2 halves =

1 whole pie

1/4 a pie

+ 1/4 a pie

+ 1/4 a pie

+ 1/4 a pie

4 fourths =

1 whole pie

1/8 a pie

+ 1/8 a pie

+ 1/8 a pie

+ 1/8 a pie

+ 1/8 a pie

+ 1/8 a pie

+ 1/8 a pie

+ 1/8 a pie

8 eighths =

1 whole pie

1/2 a pie = 1 half

+ 1/4 a pie = 1 fourth

+ 1/8 a pie = 1 eighth

+ 1/8 a pie = 1 eighth

1 whole pie

Make a circle graph to show how much pie a family ate. Here is the information you will need.

Mother ate 1/4 of the pie.

Sister ate 1/4 of the pie.

Father ate 1/4 of the pie.

Brother ate 1/8 of the pie.

Grandma ate 1/8 of the pie.

Color the graph below using the Color

Pie My Family Ate

Color Key

sister = orange   mother = pink

grandma = red   brother = yellow

father = blue

Practice 36

Directions: Circle which fraction is not in simplest form.

1. a. $\frac{1}{7}$ b. $\frac{4}{8}$ c. $\frac{2}{5}$ d. $\frac{2}{7}$

2. a. $\frac{3}{4}$ b. $\frac{2}{9}$ c. $\frac{2}{7}$ d. $\frac{3}{9}$

3. a. $\frac{5}{6}$ b. $\frac{7}{11}$ c. $\frac{2}{12}$ d. $\frac{7}{9}$

4. a. $\frac{9}{11}$ b. $\frac{5}{15}$ c. $\frac{9}{17}$ d. $\frac{10}{13}$

5. a. $\frac{6}{8}$ b. $\frac{5}{9}$ c. $\frac{1}{22}$ d. $\frac{8}{33}$

6. a. $\frac{7}{22}$ b. $\frac{5}{21}$ c. $\frac{8}{20}$ d. $\frac{5}{8}$

Directions: Write each fraction in simplest form.

7. $\frac{18}{45} = \underline{\quad}$ 10. $\frac{6}{15} = \underline{\quad}$ 13. $\frac{16}{32} = \underline{\quad}$

8. $\frac{30}{40} = \underline{\quad}$ 11. $\frac{9}{12} = \underline{\quad}$ 14. $\frac{20}{25} = \underline{\quad}$

9. $\frac{12}{14} = \underline{\quad}$ 12. $\frac{15}{18} = \underline{\quad}$ 15. $\frac{14}{28} = \underline{\quad}$

Directions: Find the least common multiple for each pair of numbers.

16. 2   18. 6   20. 4
    5   9   6

17. 3   19. 8   21. 5
    7   3   8

Directions: Circle the greater fraction in each pair.

22. $\frac{2}{5}$ $\frac{7}{15}$ 24. $\frac{5}{6}$ $\frac{3}{4}$ 26. $\frac{3}{4}$ $\frac{5}{8}$

23. $\frac{1}{2}$ $\frac{3}{8}$ 25. $\frac{2}{3}$ $\frac{4}{6}$ 27. $\frac{7}{8}$ $\frac{5}{6}$

Test Practice 1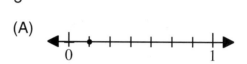

1. What fraction of the circle is *shaded*?

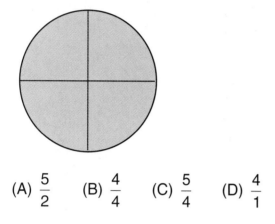

(A) $\frac{5}{2}$　　(B) $\frac{4}{4}$　　(C) $\frac{5}{4}$　　(D) $\frac{4}{1}$

2. What fraction of the rectangle is *shaded*?

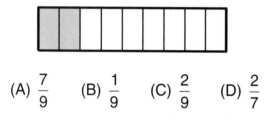

(A) $\frac{7}{9}$　　(B) $\frac{1}{9}$　　(C) $\frac{2}{9}$　　(D) $\frac{2}{7}$

3. What fraction of the circle is *shaded*?

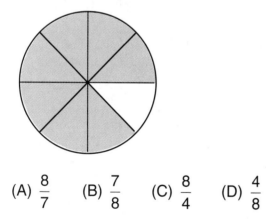

(A) $\frac{8}{7}$　　(B) $\frac{7}{8}$　　(C) $\frac{8}{4}$　　(D) $\frac{4}{8}$

4. What fraction of the rectangle is *shaded*?

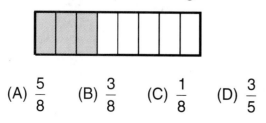

(A) $\frac{5}{8}$　　(B) $\frac{3}{8}$　　(C) $\frac{1}{8}$　　(D) $\frac{3}{5}$

5. Which number line shows the fraction $\frac{1}{5}$?

(A)

(B)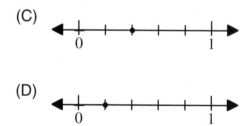

(C)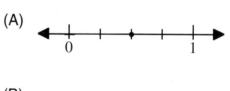

(D)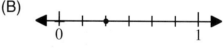

6. Which number line shows the fraction $\frac{2}{4}$?

(A)

(B)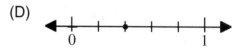

(C)

(D)

Test Practice 2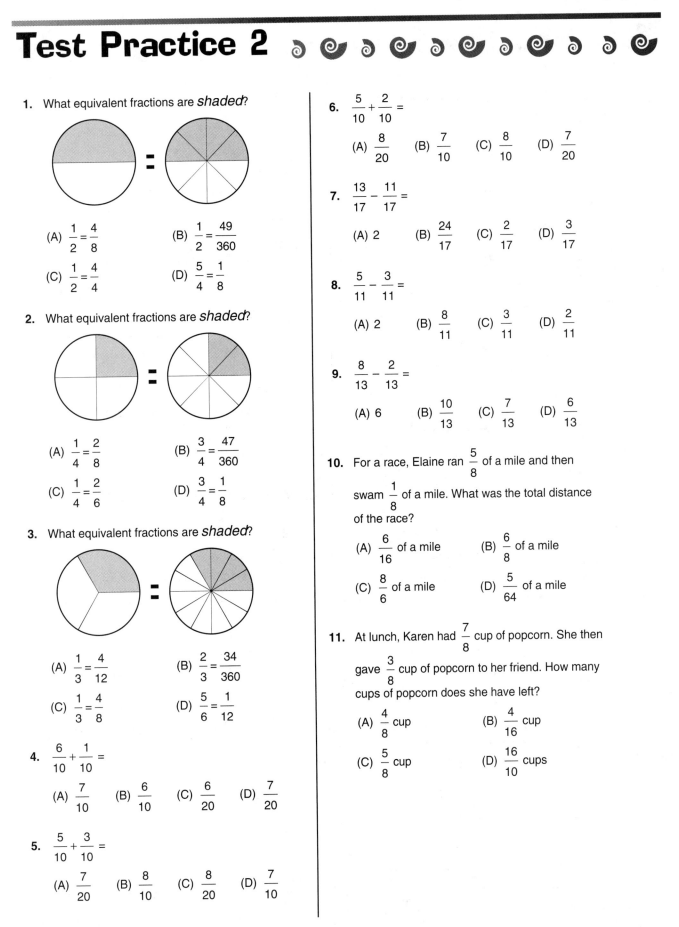

1. What equivalent fractions are *shaded*?

    (A) $\dfrac{1}{2} = \dfrac{4}{8}$ (B) $\dfrac{1}{2} = \dfrac{49}{360}$

    (C) $\dfrac{1}{2} = \dfrac{4}{4}$ (D) $\dfrac{5}{4} = \dfrac{1}{8}$

2. What equivalent fractions are *shaded*?

    (A) $\dfrac{1}{4} = \dfrac{2}{8}$ (B) $\dfrac{3}{4} = \dfrac{47}{360}$

    (C) $\dfrac{1}{4} = \dfrac{2}{6}$ (D) $\dfrac{3}{4} = \dfrac{1}{8}$

3. What equivalent fractions are *shaded*?

    (A) $\dfrac{1}{3} = \dfrac{4}{12}$ (B) $\dfrac{2}{3} = \dfrac{34}{360}$

    (C) $\dfrac{1}{3} = \dfrac{4}{8}$ (D) $\dfrac{5}{6} = \dfrac{1}{12}$

4. $\dfrac{6}{10} + \dfrac{1}{10} =$

    (A) $\dfrac{7}{10}$ (B) $\dfrac{6}{10}$ (C) $\dfrac{6}{20}$ (D) $\dfrac{7}{20}$

5. $\dfrac{5}{10} + \dfrac{3}{10} =$

    (A) $\dfrac{7}{20}$ (B) $\dfrac{8}{10}$ (C) $\dfrac{8}{20}$ (D) $\dfrac{7}{10}$

6. $\dfrac{5}{10} + \dfrac{2}{10} =$

    (A) $\dfrac{8}{20}$ (B) $\dfrac{7}{10}$ (C) $\dfrac{8}{10}$ (D) $\dfrac{7}{20}$

7. $\dfrac{13}{17} - \dfrac{11}{17} =$

    (A) 2   (B) $\dfrac{24}{17}$ (C) $\dfrac{2}{17}$ (D) $\dfrac{3}{17}$

8. $\dfrac{5}{11} - \dfrac{3}{11} =$

    (A) 2   (B) $\dfrac{8}{11}$ (C) $\dfrac{3}{11}$ (D) $\dfrac{2}{11}$

9. $\dfrac{8}{13} - \dfrac{2}{13} =$

    (A) 6   (B) $\dfrac{10}{13}$ (C) $\dfrac{7}{13}$ (D) $\dfrac{6}{13}$

10. For a race, Elaine ran $\dfrac{5}{8}$ of a mile and then swam $\dfrac{1}{8}$ of a mile. What was the total distance of the race?

    (A) $\dfrac{6}{16}$ of a mile   (B) $\dfrac{6}{8}$ of a mile

    (C) $\dfrac{8}{6}$ of a mile   (D) $\dfrac{5}{64}$ of a mile

11. At lunch, Karen had $\dfrac{7}{8}$ cup of popcorn. She then gave $\dfrac{3}{8}$ cup of popcorn to her friend. How many cups of popcorn does she have left?

    (A) $\dfrac{4}{8}$ cup   (B) $\dfrac{4}{16}$ cup

    (C) $\dfrac{5}{8}$ cup   (D) $\dfrac{16}{10}$ cups

Test Practice 3 ℰ ℯ ℯ ℯ ℯ ℯ ℯ ℯ ℯ ℯ ℯ

Directions: Fill in the circle for the correct answer to each addition problem. Choose none of these if the right answer is not given.

1.
$$\frac{1}{3}$$
$$+ \frac{1}{3}$$

(A) 2/3
(B) 2/6
(C) 1/6
(D) 1
(E) none of these

2.
$$\frac{1}{7}$$
$$+ \frac{4}{7}$$

(A) 3/7
(B) 4/7
(C) 5/7
(D) 5/14
(E) none of these

3.
$$\frac{2}{3} + 3 =$$

(A) 5 2/5
(B) 3
(C) 3 2/5
(D) 2 2/5
(E) none of these

4.
$$\frac{1}{9}$$
$$+ \frac{3}{9}$$

(A) 4/9
(B) 4/18
(C) 1 3/9
(D) 4
(E) none of these

5.
$$\frac{3}{5} + \frac{1}{5} =$$

(A) 1/5
(B) 3/5
(C) 4/5
(D) 4/10
(E) none of these

6.
$$\frac{3}{8} + \frac{2}{8} =$$

(A) 1/8
(B) 5/8
(C) 1/16
(D) 5/16
(E) none of these

7.
$$4 + \frac{1}{8} =$$

(A) 5 1/8
(B) 4 1/8
(C) 4
(D) 5/8
(E) none of these

8.
$$\frac{3}{8}$$
$$\frac{4}{8}$$
$$+$$

(A) 1/8
(B) 7/8
(C) 7/16
(D) 7
(E) none of these

9.
$$\frac{1}{2} + \frac{1}{2} =$$

(A) 1/4
(B) 2/4
(C) 3/4
(D) 1
(E) none of these

10.
$$\frac{2}{11}$$
$$+ \frac{3}{11}$$

(A) 5/11
(B) 1/11
(C) 13/14
(D) 5/22
(E) none of these

Test Practice 4 ꙮ ꙮ ꙮ ꙮ ꙮ ꙮ ꙮ ꙮ ꙮ ꙮ ꙮ

Directions: Fill in the circle for the correct answer to each subtraction problem. Choose *none of these* if the right answer is not given.

1

$$\frac{9}{11} - \frac{2}{11} =$$

(A) 2/11
(B) 5/11
(C) 7/11
(D) 9/11
(E) none of these

2.

$$\frac{3}{5} - \frac{2}{5} =$$

(A) 1/10
(B) 1/5
(C) 5/5
(D) 5/10
(E) none of these

3.

$$\frac{5}{8} - \frac{4}{8} =$$

(A) 1/8
(B) 1/16
(C) 9/8
(D) 9/16
(E) none of these

4.

$$\frac{2}{3} - \frac{1}{3} =$$

(A) 1/3
(B) 1/6
(C) 1
(D) 3
(E) none of these

5.

$$\frac{7}{12} - \frac{6}{12} =$$

(A) 1/24
(B) 1/12
(C) 13/12
(D) 13/24
(E) none of these

6.

$$\frac{11}{13}$$
$$- \frac{5}{13}$$

(A) 7/13
(B) 16/13
(C) 4/13
(D) 6/13
(E) none of these

7.

$$\frac{6}{7}$$
$$- \frac{2}{7}$$

(A) 4/7
(B) 8/7
(C) 4/14
(D) 12/14
(E) none of these

8.

$$\frac{7}{8}$$
$$- \frac{4}{8}$$

(A) 1/8
(B) 3/8
(C) 11/8
(D) 8
(E) none of these

9.

$$\frac{7}{10}$$
$$- \frac{4}{10}$$

(A) 11/10
(B) 3/7
(C) 3/10
(D) 4/10
(E) none of these

10.

$$\frac{4}{8}$$
$$- \frac{3}{8}$$

(A) 1/8
(B) 2/8
(C) 1/2
(D) 1/6
(E) none of these

Test Practice 5

Directions: Fill in the circle for the correct answer to each addition problem. Choose *none of these* if the right answer is not given.

1.

$$2\frac{2}{4}$$
$$+\ 1\frac{1}{4}$$

(A) 3 3/4
(B) 3 2/4
(C) 3 3/8
(D) 3 3/16
(E) none of these

2.

$$3\frac{3}{4}$$
$$+\ 2$$

(A) 1 3/4
(B) 5 3/4
(C) 6 3/4
(D) 5 1/4
(E) none of these

3.

$$6\frac{1}{3} + 2\frac{1}{3} =$$

(A) 4 2/3
(B) 8 2/9
(C) 8 2/3
(D) 8 2/6
(E) none of these

4.

$$2\frac{6}{7}$$
$$+\ 5$$

(A) 2 1/7
(B) 3 6/7
(C) 5 6/7
(D) 7 6/7
(E) none of these

5.

$$8\frac{1}{7} + 1\frac{2}{7} =$$

(A) 7 1/7
(B) 8 3/7
(C) 9 3/7
(D) 9 3/14
(E) none of these

6.

$$5\frac{2}{5} + 10\frac{2}{5} =$$

(A) 15 2/5
(B) 15 4/5
(C) 15 4/10
(D) 15
(E) none of these

7.

$$1\frac{1}{5} + 2\frac{1}{5} =$$

(A) 3 2/10
(B) 3 1/5
(C) 1 2/5
(D) 1 2/10
(E) none of these

8.

$$5\frac{2}{8} + 6\frac{3}{8} =$$

(A) 1 5/8
(B) 11 5/8
(C) 11 5/16
(D) 11 6/8
(E) none of these

9.

$$2\frac{3}{10} + 6\frac{4}{10} =$$

(A) 8 1/10
(B) 8 7/20
(C) 12 7/10
(D) 12 7/20
(E) none of these

10.

$$14\frac{1}{2}$$
$$+\ 1$$

(A) 14 1/2
(B) 15
(C) 15 1/2
(D) 16
(E) none of these

Test Practice 6 ➲ ✺ ➲ ✺ ➲ ✺ ➲ ✺ ➲ ✺ ➲ ➲ ✺

Directions: Fill in the circle for the correct answer to each subtraction problem. Choose *none of these* if the right answer is not given.

1.
$$6\frac{1}{2}$$
$$-\,2\frac{1}{2}$$

(A) 4
(B) 4 1/2
(C) 8 1/2
(D) 9
(E) none of these

2.
$$5\frac{2}{3}$$
$$-\,1\frac{1}{3}$$

(A) 4
(B) 4 1/6
(C) 4 1/3
(D) 4 3/6
(E) none of these

3.
$$5\frac{12}{13}$$
$$-\,4\frac{11}{13}$$

(A) 1
(B) 1 1/13
(C) 1 1/26
(D) 9 1/13
(E) none of these

4.
$$6\frac{12}{13}-5\frac{8}{13}=$$

(A) 11 4/13
(B) 4 4/13
(C) 4 13/4
(D) 1 4/13
(E) none of these

5.
$$6\frac{7}{8}-5\frac{7}{8}=$$

(A) 1 7/8
(B) 1 1/8
(C) 1 14/8
(D) 1 14/16
(E) none of these

6.
$$7\frac{7}{8}-3\frac{4}{8}=$$

(A) 3 3/8
(B) 4 3/8
(C) 4 1/8
(D) 4
(E) none of these

7.
$$17\frac{8}{11}$$
$$-\,5\frac{3}{11}$$

(A) 12 3/8
(B) 12 5/11
(C) 12 8/11
(D) 12 11/11
(E) none of these

8.
$$5\frac{4}{9}-5\frac{2}{9}=$$

(A) 5 2/9
(B) 5 2/18
(C) 6/9
(D) 2/9
(E) none of these

9.
$$8\frac{3}{4}$$
$$-\,6$$

(A) 2
(B) 2 3/4
(C) 14
(D) 14 3/4
(E) none of these

10.
$$7\frac{3}{5}$$
$$-\,5\frac{2}{5}$$

(A) 1 1/5
(B) 2 1/5
(C) 2 2/5
(D) 12 2/5
(E) none of these

Answer Sheet

Test Practice 1

1. Ⓐ Ⓑ Ⓒ Ⓓ
2. Ⓐ Ⓑ Ⓒ Ⓓ
3. Ⓐ Ⓑ Ⓒ Ⓓ
4. Ⓐ Ⓑ Ⓒ Ⓓ
5. Ⓐ Ⓑ Ⓒ Ⓓ
6. Ⓐ Ⓑ Ⓒ Ⓓ

Test Practice 2

1. Ⓐ Ⓑ Ⓒ Ⓓ
2. Ⓐ Ⓑ Ⓒ Ⓓ
3. Ⓐ Ⓑ Ⓒ Ⓓ
4. Ⓐ Ⓑ Ⓒ Ⓓ
5. Ⓐ Ⓑ Ⓒ Ⓓ
6. Ⓐ Ⓑ Ⓒ Ⓓ
7. Ⓐ Ⓑ Ⓒ Ⓓ
8. Ⓐ Ⓑ Ⓒ Ⓓ
9. Ⓐ Ⓑ Ⓒ Ⓓ
10. Ⓐ Ⓑ Ⓒ Ⓓ
11. Ⓐ Ⓑ Ⓒ Ⓓ

Test Practice 3

1. Ⓐ Ⓑ Ⓒ Ⓓ Ⓔ
2. Ⓐ Ⓑ Ⓒ Ⓓ Ⓔ
3. Ⓐ Ⓑ Ⓒ Ⓓ Ⓔ
4. Ⓐ Ⓑ Ⓒ Ⓓ Ⓔ
5. Ⓐ Ⓑ Ⓒ Ⓓ Ⓔ
6. Ⓐ Ⓑ Ⓒ Ⓓ Ⓔ
7. Ⓐ Ⓑ Ⓒ Ⓓ Ⓔ
8. Ⓐ Ⓑ Ⓒ Ⓓ Ⓔ
9. Ⓐ Ⓑ Ⓒ Ⓓ Ⓔ
10. Ⓐ Ⓑ Ⓒ Ⓓ Ⓔ

Test Practice 4

1. Ⓐ Ⓑ Ⓒ Ⓓ Ⓔ
2. Ⓐ Ⓑ Ⓒ Ⓓ Ⓔ
3. Ⓐ Ⓑ Ⓒ Ⓓ Ⓔ
4. Ⓐ Ⓑ Ⓒ Ⓓ Ⓔ
5. Ⓐ Ⓑ Ⓒ Ⓓ Ⓔ
6. Ⓐ Ⓑ Ⓒ Ⓓ Ⓔ
7. Ⓐ Ⓑ Ⓒ Ⓓ Ⓔ
8. Ⓐ Ⓑ Ⓒ Ⓓ Ⓔ
9. Ⓐ Ⓑ Ⓒ Ⓓ Ⓔ
10. Ⓐ Ⓑ Ⓒ Ⓓ Ⓔ

Test Practice 5

1. Ⓐ Ⓑ Ⓒ Ⓓ Ⓔ
2. Ⓐ Ⓑ Ⓒ Ⓓ Ⓔ
3. Ⓐ Ⓑ Ⓒ Ⓓ Ⓔ
4. Ⓐ Ⓑ Ⓒ Ⓓ Ⓔ
5. Ⓐ Ⓑ Ⓒ Ⓓ Ⓔ
6. Ⓐ Ⓑ Ⓒ Ⓓ Ⓔ
7. Ⓐ Ⓑ Ⓒ Ⓓ Ⓔ
8. Ⓐ Ⓑ Ⓒ Ⓓ Ⓔ
9. Ⓐ Ⓑ Ⓒ Ⓓ Ⓔ
10. Ⓐ Ⓑ Ⓒ Ⓓ Ⓔ

Test Practice 6

1. Ⓐ Ⓑ Ⓒ Ⓓ Ⓔ
2. Ⓐ Ⓑ Ⓒ Ⓓ Ⓔ
3. Ⓐ Ⓑ Ⓒ Ⓓ Ⓔ
4. Ⓐ Ⓑ Ⓒ Ⓓ Ⓔ
5. Ⓐ Ⓑ Ⓒ Ⓓ Ⓔ
6. Ⓐ Ⓑ Ⓒ Ⓓ Ⓔ
7. Ⓐ Ⓑ Ⓒ Ⓓ Ⓔ
8. Ⓐ Ⓑ Ⓒ Ⓓ Ⓔ
9. Ⓐ Ⓑ Ⓒ Ⓓ Ⓔ
10. Ⓐ Ⓑ Ⓒ Ⓓ Ⓔ

Answer Key

Page 4
1. 1/3
2. 1/4
3. 5/6
4. 3/5
5. 7/10
6. 2/6 or 1/3
7. 3/4
8. 1/2

Page 5
1. 1/3
2. 4/6 or 2/3
3. 2/5
4. 3/4
5. 1/2
6. 1/2
7. 5/9
8. 1/4
9. 3/4
10. 3/6 or 1/2
11. 2/4 or 1/2
12. 2/3

Page 6
Order of fractions:
1/2, 1/3, 1/4, 1/5, 1/6,
1/7, 1/8, 1/9, 1/10

Page 7
1. D
2. B
3. B
4. C
5. C
6. B

Page 8
1. 1/2
2. 2/3
3. 2/4 or 1/2
4. 1/2 < 3/4
5. 1/4 < 1/2
6. 3/8
7. 4/8 or 1/2
8. 5/8
9. 2/5 < 3/5
10. 2/4 > 1/4

Page 9
1. A
2. A
3. A
4. D
5. B
6. B
7. A
8. D

Page 10
1. D
2. C

3. A
4. D
5. A
6. B
7. B
8. A

Page 11
1. 5/6
2. 2/6
3. 1/8
4. 1/4
5. 4/7
6. 1/6
7. 4/8

Page 12
1. 1/2
2. 1/7
3. 1/9
4. 5
5. 3
6. 1/6
7. 1/3
8. 1/1
9. 8
10. 6

Page 13
1. C
2. C
3. A
4. B
5. D
6. B

Page 14
1. 2/7, 5/7, 6/7, 7/7
2. 1/11, 3/11, 5/11, 11/11
3. 1/10 <1/4
4. 2/13, 3/13, 8/13, 13/13
5. 1/5, 2/5, 3/5, 5/5
6. 1/4 > 1/8
7. 2/11, 7/11, 9/11, 11/11
8. 2/13, 8/13, 12/13, 13/13
9. 1/3 > 1/6
10. 1/5, 2/5, 3/5, 5/5
11. 1/6 < 1/3
12. 2/5, 3/5, 4/5, 5/5
13. 3/7, 4/7, 5/7, 7/7
14. 2/13, 4/13, 12/13, 13/13
15. 1/8 = 1/8

Page 15
1. 1/4, 2/4, 3/4, 1
2. 1/3, 2/3, 1

3. 1/5, 2/5, 3/5, 4/5, 1
4. circle second circle, underline first circle
5. circle third circle, underline second circle
6. 5/6
7. 7/8
8. 2/4
9. 15
10. 20

Page 16
1. D
2. B
3. D
4. C
5. B
6. C

Page 17
1. shade 3 parts, 3/6
2. shade 4 parts, 4/6
3. shade 6 parts, 6/8
4. shade 2 parts, shade 3 parts, 2/6, 3/9
5. shade 4 parts, shade 2 parts, 4/8, 2/4
6. 2/4
7. 4/6
8. 4/8
9. 2/4, 4/8
10. 2/6, 3/9

Page 18
1. b
2. a
3. b
4. b
5. c
6. 3, 9
7. 2, 8
8. 6, 12
9. 2, 4
10. 3, 3
11. 8, 1
12. 2, 5
13. a
14. c

Page 19
1. 3/4
2. 2/5

3. 1/3
4. 1/2
5. 1/2
6. 1/4
7. 1/3
8. 1/4
9. 1/3
10. 1/5
11. 1/2
12. 1/7
13. yes
14. no
15. yes
16. yes
17. no
18. yes
19. yes
20. yes
21. yes
22. no
23. no
24. no

Page 20
1. 3/4
2. 1/2
3. 1/2
4. 1/3
5. 2/3
6. 5/10
7. 2/12
8. 6/8
9. Answers will vary.
10. Answers will vary.

Page 21
1. D
2. B
3. A
4. B
5. C
6. B
7. A
8. B
9. D
10. C

Page 22
1. 1
2. 2/5
3. 3/8
4. 7/9
5. 7/20
6. 3/7
7. 4/5
8. 1
9. 17/24

3. 1/3
4. 1/2
5. 1/2
6. 1/4
7. 1/3
8. 1/4
9. 1/3
10. 1/5
11. 1/2
12. 1/7
13. yes
14. no
15. yes
16. yes
17. no
18. yes
19. yes
20. yes
21. yes
22. no
23. no
24. no

10. 5/7
11. 4/5
12. 7/8
13. 7/10
14. 2/3
15. 11/13
16. 2/3

Page 23
1. 6/7
2. 4/5
3. 7/12
4. 7/8
5. 7/9
6. 11/14
7. 3/5
8. 11/17
9. 1/3
10. 7/13
11. 7/8
12. 15/19
13. 3/4
14. 13/16
15. 1
16. 7/10

Page 24
1. 3
2. 3 2/3
3. 7 1/2
4. 1 1/2
5. 6 1/3
6. 5
7. 5 3/4
8. 11
9. 3 1/2
10. 8
11. 3 3/8
12. 8 4/7
13. 5 3/8
14. 15
15. 23
16. 134 1/2

Page 25
1. B
2. A
3. D
4. D
5. D
6. B

Page 26
1. 14 3/5
2. 19 5/7
3. 19 3/8
4. 24 3/5
5. 16 1/8
6. 11 3/5
7. 10 1/8

Answer Key

Page 27
1. 2 1/2
2. 1 1/3
3. 3 1/4
4. 2 2/5
5. 1 1/6
6. 1 1/9
7. 1 2/7
8. 1 1/3
9. 1 3/7
10. 1 3/5
11. 2 1/4
12. 1 1/2
13. 1 2/3
14. 2 1/2
15. 1 5/9
16. 2 2/3
17. 1 3/4
18. 3 2/3
19. 1 1/3
20. 1 8/9

Page 28
1. B
2. C
3. A
4. D
5. C
6. D
7. A
8. C

Page 29
1. 1/12
2. 5/8
3. 3/8
4. 1/3
5. 2/3
6. 0
7. 4/5
8. 3/8
9. 1/2
10. 1/5
11. 8/11
12. 3/5
13. 1/10
14. 1/4
15. 1/3
16. 1/14

Page 30
1. 2/5
2. 2/13
3. 2/13
4. 1/3
5. 1/2
6. 7/10
7. 7/9
8. 3/7
9. 1/2
10. 3/11
11. 3/16
12. 3/7
13. 1/10
14. 0
15. 7/17
16. 3/19

Page 31
1. 1/8
2. 1/6
3. 5/7
4. 1 1/3
5. 1 1/4
6. 3/7
7. 3/8
8. 1/4
9. 2/4 or 1/2
10. 2/6 or 1/3

Page 32
1. 2 1/2
2. 7 5/8
3. 9 1/3
4. 4 7/9
5. 1 1/2
6. 12 7/10
7. 2 2/5
8. 4 3/4
9. 5 1/7
10. 13 6/17
11. 4 1/3
12. 7 11/21
13. 2 7/13
14. 13 4/7
15. 8 1/12
16. 28 2/11

Page 33
1. 9/100
2. 44/100
3. 98/100
4. 65/100
5. 39/100
6. 27/100
7. 73/100
8. 88/100
9. 23/100
10. 15/100
11. 50/100
12. 11/100
13. 46 < 99/100
14. 57 < 83/100
15. 18¢ = 18/100
16. $0.25 > 20/100
17. $0.63 = 63/100
18. $0.74 > 31/100

Page 34
1. 5 81/100
2. 2 71/100
3. 3 19/100
4. 1 86/100
5. 4 4/100
6. 5 11/100
7. 1 63/100
8. 2 10/100
9. 3 7/100
10. 4 29/100
11. 7 99/100
12. 3 0/100 or 3
13. $6.19
14. $2.01
15. $4.98
16. $3.68
17. $1.50
18. $2.27

Page 35
1. 3 3/4
2. 2 1/4
3. 1 1/5
4. 1 2/7, 4 2/3
5. 3 1/2, 1 1/5
6. 2/3
7. 7/8
8. 5/6
9. 3/4
10. 4/5

Page 36
1. 0
2. 4/9
3. 3/7
4. 3/8
5. 11/24
6. 3
7. 2
8. 1
9. 3, 3
10. 2, 2

Page 37
1. 5
2. 4
3. 4
4. 2, 5
5. 5, 12
6. 3/10
7. 4/5
8. 2/3
9. The spinner should have 3 blue parts and 2 red parts.
10. The spinner should have 2 blue parts and 3 red parts.

Page 38

Page 39
1. b
2. d
3. c
4. b
5. a
6. c
7. 2/5
8. 3/4
9. 6/7
10. 2/5
11. 3/4
12. 5/6
13. 1/2
14. 4/5
15. 1/2
16. 10
17. 21
18. 18
19. 24
20. 12
21. 40
22. 7/15
23. 1/2
24. 5/6
25. They are equal.
26. 3/4
27. 7/8

Page 40
1. B
2. C
3. B
4. B
5. D
6. A

Page 41
1. A
2. A
3. A
4. A
5. B
6. B
7. C
8. D
9. D
10. B
11. A

Page 42
1. A
2. C
3. E
4. A
5. C
6. B
7. B
8. B
9. D
10. A

Page 43
1. C
2. B
3. A
4. A
5. B
6. B
7. A
8. B
9. C
10. A

Page 44
1. A
2. B
3. C
4. D
5. C
6. B
7. E
8. B
9. E
10. C

Page 45
1. A
2. C
3. B
4. D
5. E
6. B
7. B
8. D
9. B
10. B